Buffoon

Ken Coffman

Other books by Ken Coffman

Steel Waters
Alligator Alley (with Mark Bothum)
Twisted Shadow (with Mark Bothum)
Glen Wilson's Bad Medicine
Toxic Shock Syndrome
Immortality, LLC
Hartz String Theory
Endangered Species
Fairhaven
Mesh (with Adina Pelle)
The Sandcastles of Irakkistan
Fiona and the Black Faerie Prince (with Kristen Poeraatmadja)
Real World FPGA Design with Verilog

Buffoon: One Man's Cheerful Interaction with the Harbingers of Global Warming Doom
©2014 Ken Coffman
All Rights Reserved

ISBN 978-1-941071-03-8

Cover Design by Guy Corp, www.grafixCORP.com
Front Cover Illustration by Josh, CartoonsByJosh.com

STAIRWAY⹀PRESS

SEATTLE

www.stairwaypress.com
1500A East College Way #554
Mount Vernon, WA 98273
(360) 336-3366
Contact: Stacey@StairwayPress.com

There are very few (if any) mentions of the Sky Dragon Slayers and PSI (Principia Scientific International, www.principia-scientific.org) in this volume. Why? Because this volume grew too long to print in a cost-effective manner, so I'm saving the events surrounding the publication of their book—and the heated discussions that followed—for the second volume where the Sky Dragon Slayers will play a central role.

Very few of the ideas discussed in this book are original. As you'll conclude from looking at the bibliography and this book's companion disc, I immersed myself in the topic of human-caused global warming and consumed papers, blog posts and commentary in mass quantities. I give all due credit to all the bloggers and the active online community, but many wanted nothing to do with this book, so I omitted many proper references. Herein, should you accidentally stumble over any interesting facts, clever analysis or deep thoughts, they were probably conceived by someone else. To all the people who filled my head with information and guided my thinking, I salute you.

I believe the battles, except for mopping up and marginalizing key militants who will cling to faulty, self-serving memes to their deathbeds, are over. The Sky Dragon Slayers are the target of much derision, but as soon as the effect of atmospheric CO_2 is recognized as tiny and immeasurable—and the error bands in the data properly include the possibility that CO_2 is a coolant, then everyone will be a Sky Dragon Slayer, though I doubt we'll ever get proper credit. So it goes. It was a fun time to be skeptical of progressive nonsense.

I hope you enjoy my happy-go-lucky adventures.

DEDICATION

While flying from Hong Kong to Penang, Malaysia, I struck up a conversation with my seatmates—as you might imagine, this is something I often do. My neighbors were two Russian scholars specializing in catalytic chemistry. There was no way I would neglect asking their opinion about the science of man-made global warming. The young lady by the window leaned over and in a charming Russian accent voiced the most succinct debunking of the global warming theory I've heard. The degree of nonsense endorsed by climate activists cannot be made more plain or stated more concisely than this:

There is no mechanism.

I dedicate this book to Maria Bykova from Novosibirsk, Russia— may she live long and prosper.

AN OPENING ROUND

The Heart of the Matter

I've just completed Mike's Nature trick of adding in the real temps to each series for the last 20 years (ie from 1981 onwards) and from 1961 for Keith's to hide the decline.
—Phil Jones, PhD, Director of Research, University of East Anglia, Climatic Research Unit

THE TITLE OF this book is inspired by inspiring words of wisdom from one of my biggest fans:

DickWad: *I can see why the RealClimate guys consider you to be a buffoon with a fourth grade level understanding of climate science.*

Ah, aren't the climate activists sweet? All they want to do is save the world. If only recalcitrant folks like me would simply get out of the way and let them get on with their important business.

You don't have to waste your time wading through the entirety of this rambling, juvenile excuse for a book to get to the central conclusion I want to communicate. I'll cough it up right here and make it easy on you. Here it is:

There are people who will lie to you—directly and to your face. That is simply a fact. In addition, there are people who are dead wrong and will try to convince you to embrace the most

1

absurd, farfetched twaddle if it somehow benefits them. In general, as a species, our level of self-deception and wrong-headed ignorance is remarkable and examples are endless[1], just engage your brain and listen to the radio, watch TV or surf the Internet for five minutes. Let the buyer beware (caveat emptor), indeed.

So, how do we, as intelligent, aware people, navigate this endless sea of human nonsense?

Regardless of what names you are called (buffoon!) or how much your opponent makes you feel stupid for not embracing their brilliance, stand your ground. Though your opponents might have PhDs, lots of like-minded friends or look great in fancy suits, don't assume they know what they are talking about or are telling the truth, particularly if they have a vested interest. People will lie and expect you to eat it up and ask for more; this is a central tenet of human nature.

Here's something to watch for: the nature of thermodynamics and atmospheric physics, in detail, is complex, but the general concepts are simple—easily grasped and conceived—and don't let anyone tell you otherwise. I'm telling you straight up, if a buffoon like me can understand the fundamentals (and I do), then *you* can too. Don't let the activists convince you this science is difficult to understand—it doesn't take massive horsepower between the ears to grasp facts like this: to warm something, you need generally need

---

[1] Everyone should have a copy of Charles MacKay's *Extraordinary Popular Delusions*, first published in 1841. You can download it for free from Gutenberg.org and other places.

In reading the history of nations, we find that, like individuals, they have their whims and their peculiarities; their seasons of excitement and recklessness, when they care not what they do. We find that whole communities suddenly fix their minds upon one object, and go mad in its pursuit; that millions of people become simultaneously impressed with one delusion, and run after it, till their attention is caught by some new folly more captivating than the first.

something warmer. There are exceptions, of course, like exothermic chemical or nuclear reactions and heating via microwaves, but I haven't heard any climate scientists suggest these methods are at work in our atmosphere.

You don't have to know the formal definition of thermal mass to intuitively know why lake water or firebricks retain heat after the Sun goes down (demonstrating the relative thermal capacity of water and brick compared to air). You don't need to be a rocket scientist to understand that passive radiation is a weak method of moving energy around—and is ineffectual unless it comes from an intense source like the Sun, a fire or a nuclear bomb.

You've heard the expression: it's not the heat, it's the humidity? You don't have to be a certified expert in evaporative cooling to understand the difference between 105°F (40.5°C) afternoons in Phoenix, Arizona (dry) and Orlando, Florida (humid). Your body cools itself via sweat evaporation—if the air is humid, then evaporation is reduced and you feel warmer. That's simple enough, right?

Have you noticed this? Any honest educator or expert who fully understands a subject can describe the central tenets of their topic of expertise in simple and easy-to-understand terms. Obfuscation is a signal of deceit—that someone is trying to convince you to believe a line of bullshit.

I'm not saying you should be willfully stupid. Gather facts. Educate yourself. Think things through and trust your family, friends, instincts, intellect and intuition.

Here is a hoary cliché immortalized on bumper stickers: Eschew Obfuscation. It means unnecessary complexity is not a virtue. In fact, let's capture this truism as Coffman's Law:

The more complex an explanation, the more likely it's complete nonsense.
—Coffman's Law

As an illustration, think of the financial derivatives that were a factor in the worldwide financial meltdown that started in 2006—where a few trillion dollars evaporated almost overnight. Credit Derivatives: blah-blah-blah. Like my grandfather said: *When a politician comes to the door: keep your hand on your wallet.*

Mother Nature can be kind to intuitive thinkers. For most of the things we need to know in life, there are simple, common sense analogies that can be used. The analogy might not be rigorous and detailed, but will serve a useful purpose. Here are some examples:

- *What goes up must come down.* Gravity: it's the law. It sucks—there is a relentless force pulling objects toward the center of the Earth.

- *The taller they stand, the harder they fall.* Leverage. Center of gravity. Balance. These are all important, easy-to-grasp physical properties.

- *Don't put the cart before the horse.* Unless you're a climate scientist, cause comes before effect, right?

- *Everybody talks about the weather, but no one does anything about it.* It is the ultimate hubris to think that humans running around on the 14.5% of the Earth's surface that is useful to us can affect the weather or climate. We're just not that powerful, my friends. We couldn't do it on purpose, so you think we can do it by accident? Think again.

- *Exceptions prove the rule.* For all generalizations, there will be exceptions. In fact, you should never generalize (ha!).

- *Follow the money.* Greed is a key element of our human nature. People will lie, cheat and steal to get easy money. Sadly, people will even lie to themselves to get ahead; we're pathetic creatures capable of rationalizing anything.

- *Garbage in-garbage out.* If a model, algorithm or formula uses bad (unconstrained, noisy, inappropriate or illegitimate) data, then the result cannot be trusted.

- *Good enough for government work.* Only competition keeps any of us honest—all monopolies are inefficient. If you want high quality results, don't seek them in a bureaucratic organization.

- *If it walks like a duck and quacks like a duck—it's a duck.* Don't let anyone convince you that what you see is wrong. Observations trump theory.

- *Keep It Simple, Stupid (KISS).* This rephrasing of Occam's razor says we should pick the competing hypothesis that makes the fewest assumptions and thereby offers the simplest explanation of an effect.

- *Lightning never strikes the same place twice.* This is a comment about the vastness and randomness of the universe.

- *March goes in like a lion, but out like a lamb.* This is a generalization about the weather, though your mileage on this truism might vary in the Southern hemisphere. That's climate science humor, my friends.

- *Nature abhors a vacuum.* To be more general, Mother Nature likes things fully averaged out with matter and energy fully integrated and distributed and (again, excluding external work) by and large relentlessly works toward this end.

- *Never ask a barber if you need a haircut.* This goes back to vested interest—never ask a grant-seeking academic if more government funding is needed for climate research. Never ask a progressive climatologist to peer-review a paper critical of the human-caused global warming theme. Never ask a government bureaucrat if more rules and regulations are needed. Never ask a community organizing activist if more government programs are needed.

- *Never stare directly at the Sun.* This is a message about relative radiation strength. We enjoy the use of sunlight all day long if it's diffused or scattered, but staring directly at the Sun (or at a strong reflection) is a hazard. We need sunlight to be diluted to be useful for our eyeballs.

- ***Nothing lasts forever.*** This is a simplification of the Second Law of Thermodynamics[2].

- ***Numbers don't lie, but statisticians do.*** For various reasons, people project meaning into things; this is human nature. We project order onto chaos. Here are two more truisms: *Statistics are for those who believe in nothing, but wish to act on something and there are lies, damn lies, and then there's statistics*[3].

- ***One thing leads to another.*** As Isaac Newton described his clockwork universe: application of a non-zero net force on a body is both a necessary and sufficient condition for either generating or altering its true motion[4]. As I said earlier, cause first, then affect.

- ***Opposites attract.*** This generalization describes things we commonly observe including the attraction between north and south magnetic poles and electrons moving between the positive and negative terminals of a battery.

- ***Lecturer: one with his hand in your pocket, his tongue in your ear, and his faith in your patience.***[5] We'll talk more about this later, but oh, how I have come to despise many in the academic world.

If you need any further encouragement to tackle the physics of

---

[2] There are various definitions. Here's a useful one: for an isolated system, the natural course of events takes the system to a more disordered (higher entropy) state. Source: http://hyperphysics.phy-astr.gsu.edu/hbase/thermo/seclaw.html

[3] Mark Twain attributed this quote to Benjamin Disraeli.

[4] Rynasiewicz, Robert, *Newton's Views on Space, Time, and Motion*, The Stanford Encyclopedia of Philosophy (Fall 2011 Edition), Edward N. Zalta (ed.), http://plato.stanford.edu/entries/newton-stm/

[5] Ambrose Bierce, *The Devil's Dictionary*

thermodynamics and climate, let's turn to Dr. Einstein:

Imagination is more important than science[6].
—Albert Einstein

You're an imaginative person, aren't you?

There is No Evidence

I remember my shock when Don Easterbrook told me there is no evidence to support the claim of human-caused global warming[7]. At that time I was unaware of the fully corrupt nature of post-normal climate science. I assumed there was evidence, albeit weak, arguable and unsupportive of the Anthropogenic Global Warming (AGW, where anthropogenic means 'caused by man') or CAGW (Catastrophic AGW) assertions.

I am educated and trained as an engineer, a trade that requires solid links to physical reality. If I design flight-critical avionics, the design has to be based on verifiable physics; it must be proved to work and be thoroughly tested and verified in harsh environmental conditions. I am also a novelist, so I'm also comfortable in the world of simply making things up—the world of bullshit.

If we're going to rework our economy and spend literally trillions of public sector dollars on *green* technology, surely there's some evidence. There must be, right? You'd think CAGW themes aren't all based on stupid, wrong-headed models and hand-waving, but it is. Don't take my word for it, look for yourself.

Where do disturbing conclusions about mankind destroying

---

[6] *Cosmic Religion: With Other Opinions and Aphorisms* (1931) by Albert Einstein, p. 97.
[7] The question I have for people [who believe in human-caused global warming], 'please tell me what is the real physical evidence that CO_2 is the cause of AGW?' Tell me. There isn't any.
—Dr. Donald J. Easterbrook, Professor Emeritus of Geology, Western Washington University

the Earth come from? Models. All the way down, alarmism comes from models. How good are the models? If they are so great, why can't they predict next year's weather? Instead of using one great one, why are ensembles of model outputs combined to get the consensus results? Where is the proof that the various programmed control knobs (like CO_2 concentration) represent anything real in our atmosphere?

In this book, I am not telling you what to think. I'm telling you what I think and explaining why. Think for yourself and make up your own mind about what the facts tell you. There's one thing I can tell you for certain: the methods of modern climate science have nothing to do with the methods used to design things that actually have to work, like bridges and airplanes.

NOW LET'S NOT FOCUS ON HOW HOPELESS THE MODELS ARE, OK?

HORROR SCOPES by me

HORRORSCOPE HANSEN
a model of discretion

The Nature of the Difficulty is the Difficulty of Nature

To grasp some of the problem, let's think about doing something much more trivial and easy than figuring out the Earth's average temperature.

Suppose I pointed a gun to your head and threatened to shoot you unless you told me what yesterday's average temperature at your house was—within a tenth of a degree Celsius. After a moment of thought, a competent engineer might grab the gun and pull the trigger himself to end the torture.

Why?

Because this question is unanswerable—first because I have not defined the question clearly and second, because there is no data available to support this level of accuracy. The first things an analytical thinker considers is what do we know, how do we know it and what are the limits of our knowledge?

What was the average temperature outside *your* house yesterday? Perhaps you have an outdoor thermometer, but how many times did you look at it? Is it part of a system that automatically collects temperature data?

What is its accuracy? If it's decent, it will be accurate to within a degree or two. What is the state of its calibration? Has it drifted over time? Is it sited well enough to represent the average temperature outside your house—or is it exposed to direct sunlight for part of the day which will distort the data? Are there external heat sources like dryer or heater vents polluting the readings?

And, what is meant by average? Is it the average of the peak temperature in the afternoon and valley temperature early in the morning, or to be meaningful, do we need greater resolution? Shall we take twenty-four hourly readings and average them? Shall we take readings every six minutes and average the 240 readings per day? How accurate is our time sampling?

How fast does the temperature change? To be accurate, we

need to recognize the Nyquist[8] sampling rate with relation to the rate of temperature change. What sampling rate do we need to get 0.1°C accuracy? 6X the rate of temperature change? 10X? Perhaps we should use something with a large thermal mass (like a bucket of water) to integrate the temperature changes. How big should the bucket be? How often should we replenish the water to account for evaporation or the neighbor's dog coming over for a drink? Where should we put the bucket? In the shade? In direct sunlight? What site best represents the average temperature outside our house?

I bring all this up not to drive you mad with factors you might not have thought of, but to illustrate how complicated a simple problem can be. Now imagine a climate scientist telling you he or she knows the Earth's average temperature in 1900 within 1°C. Really? Do you buy it?

Do you know what flimflam artists and charlatans like? They like noisy data and uncertainty. As an example, take the street huckster hiding a pea under three walnut shells. He doesn't care about the actual location of the pea, all he cares about is randomizing the location so your guess has a one-in-three chance of being right. They make good money paying out 2:1 when the odds against the sucker are 3:1.

With randomness and uncertainty, anything can be proved. What signal do you want to extract from noisy data? With selective bias and proper filtering, hell, we could get anything we want. Want to see hockey sticks? No problem. Sometimes this is unintentional. You can *know* a signal is there and then find it without realizing how much you are projecting into it (this is called confirmation bias). Sometimes it's easier to fool yourself than to fool others.

---

[8] In essence, the theorem shows that a bandlimited analog signal can be perfectly reconstructed from an infinite sequence of samples if the sampling rate exceeds $2B$ samples per second, where B is the highest frequency of the original signal. Source: Wikipedia. The Nyquist rate is necessary, but not always sufficient. For complex signals, we generally want much higher sampling rates than the Nyquist limit.

If we look only at history and ignore hysteria derived from nonsensical models, what are we alarmed about? How much of a historic temperature change are we talking about during the period where humans could have been an influence? A generally accepted number is about 1°C or 1.6°F in the last 150 years.

Yes, that's it.

From this small change springs all the global warming alarmism. We're not very sure about the magnitude of this number because it is largely based on readings from airports which exaggerate the Urban Heat Island[9] (UHI) effect.

In addition, we don't know how much of this is simply noise—random variation from natural causes. And, is this increased heat a hazard? When you notice how crowded and popular cities are, you might conclude we like our living environment warmer than the countryside. How else can we explain the popularity of city life over rural?

I believe the Earth has slowly warmed a little over the last 150 years and I see no evidence this increase has been harmful to the human race or that further warming would not be even better. You do not want to trade today's climate with that of the 1700's. You do not want to trade today's climate for that of 20,000 years ago. Don't believe me? Check for yourself what Mother Nature did to the Earth in recent history. Want to really be scared? Google Younger Dryas stadial (The Big Freeze) and imagine how much fun it would be to live during a dramatic series of events like that.

More warmth? Bring it on.

Preconceived Notions

Let me share my preconceived notions and biases. I am inclined toward prejudging people based on my life experiences. Isn't that

---

[9] UHI. If you pay attention to the thermometer in your car, you've probably noticed urban (city) environments are generally warmer than the surrounding countryside. Why? It's due to decreased albedo of buildings and asphalt compared to foliage—and waste heat from urban activities.

the essence of wisdom? Hence, I don't like insurance agents, time share sales professionals, politicians, TV preachers, screaming kids in expensive restaurants, people who throw lighted cigarettes out of car windows, cherry-picking cops in speed traps, financial planners, stock brokers, derivative and credit default swappers, well-fed bums with cardboard signs, community organizers, tax collectors, poverty pimps, parasitic freeloaders, ivory tower academics, progressive lefties and morons who believe humans have a measurable impact on the climate.

Over the past few years, I developed a particular aversion to academics. I used to think they were silly and harmless—simply a waste of flesh, but, from reading endless climate-related papers, my opinion declined and hardened. Professors in hard sciences like engineering are fine; they routinely test their ideas against reality and can't get too deranged.

However, there is a certain type attracted to and nurtured by soft-topic academia—liberal arts like Women's Studies[10] and squishy sciences like Climatology. Because humanities and soft-science studies are subjective and friendlier to mediocre and corrupt intellects, these fields overflow with blowhards, dimwits and exploiters. I despise them for being disconnected from objective reality—they truly believe an interesting theory lining up reasonably with cherry-picked, altered data makes something real. They believe inventing a cohesive storyline is sufficient for proof. I find this offensive.

Let me say that I don't have a bias against all stupid people. If a dumb person *acts* intelligently, then that's a kind of Turing test[11], isn't it? If a smart person does *A* and a not-so-smart person does *A*, that's good enough for me. I'm no mind reader. I can't tell the difference, so, to me, both people have equal merit. I don't care if

[10] Do we really need more people with brains infected with Queer Theories and Womanist Spiritual Activism taking publically subsidized classes like Holistic Pedagogies Colloquium?
[11] http://loebner.net/Prizef/TuringArticle.html

you have an IQ of 181[12]; if you smoke crack, kick puppies or believe humans can influence the global climate, then you're stupid and I don't like you. If you have an IQ of 80 and live a good, clean wholesome life, then I like you. I respect you. Come over to the house and we'll have a microbrew and a hamburger.

Humans are storytelling, pattern-seeking animals. We project irrational meaning into chaos all the time. As an example, dare I mention Astrology? Don't the climate activists realize how cheap and sleazy correlations are as compared to root causes—which can be devilishly difficult? If you want to be a rational person, then you have to constantly fight against the impulse to project unwarranted meaning onto chaos.

I don't believe anything that is not double-blind tested. In addition, I only believe in things that can be measured. I say that to be a little provocative, but this philosophy keeps me out of trouble when it comes to Astrology, radio and TV sales pitches for get-rich schemes, flowery promises from politicians, hair restoration creams and come-ons from lovely ladies who send me email love notes, but never seem to spell my name correctly.

CAGW prophets of doom like James E. Hansen of NASA's Goddard Institute for Space Studies (GISS) are financial rainmakers attracting government funding and grants with stellar prose like this:

> What we have shown in this paper is that time is rapidly running out. The era of doubts, delays and denial, of ineffectual half-measures, must end. The period of

---

[12] 181 is chess grandmaster Bobby Fischer's estimated IQ—one point higher than actor James Woods. How smart is James Woods? Here's a quote from the National Enquirer: *Tough-guy actor James Woods' gorgeous young girlfriend—who's 39 years his junior—is crushed that the star is refusing to marry her, say sources. Blonde beauty Ashley Madison was heartbroken after the 64-year-old 'Shark' star failed to cough up and engagement ring for her 25th birthday recently, a friend of the couple told The Enquirer. Source: http://celebslam.celebuzz.com/2011/06/james-woods-girlfriend-bikini.php*

consequences is beginning. If we fail to stand up now and demand a change of course, the blame will fall on us, the current generation of adults. Our parents did not know that their actions could harm future generations. We will only be able to pretend that we did not know. And that is unforgivable.[13]

Woe, despair and calamity. What soul-wrenching piousness—oh, how they earnestly labor to save us from ourselves.

Are there any predictions we can use to decide if Dr. Hansen has the slightest clue about what he's talking about? Generally, they are careful not to say anything concrete; they stick to generalizations and spongy timelines. However, in 1988, Hansen predicted the street outside his office in Manhattan would be under water by 2030[14]. Let's not take a chance on saying anything inaccurate or out of context and study the record verbatim.

> In fact back in New York, later, one of the reporters, visiting Hansen's office, would bring the scientist to his window, and the men would gaze down at the "Flowers By Valli" shop and the green awning of Café 112 across Broadway, and the reporter would ask the scientist whether, if he was right, and carbon dioxide in the atmosphere really doubled, anything down there would look different because of it by 2030. Would average people notice any changes?
>
> Hansen would say, "There will be more traffic on

[13] http://pubs.giss.nasa.gov/docs/notyet/submitted_Hansen_etal.pdf

[14] Actually, the question was asked in 1998 with regard to a 40 year prediction, so this flood date should be 2028, not 2030. Even worse, the climate alarmists say the effect of doubling CO_2 is exponential, so most of the hazard should be present now considering the increase of CO_2 in 1988 (from 350PPM to 390PPM). According to Warmist doctrine, we should see more affect from the first 20PPM increase rather than the later 20PPM.

Broadway."

"Why?" the puzzled reporter would ask, thinking at first that Hansen had not heard the question right.

"Because the West Side Highway will be underwater...you might see Dutch engineers down there, to build dikes. The Dutch could sell their expertise in building dikes in New York, Florida, Louisiana."

"Dikes," the reporter would repeat, stunned.

"As the warming progresses," Hansen would continue, "and droughts get more severe, you might see signs in restaurants, 'Water by Request Only.' Hurricanes and thunderstorms will be more frequent. You might see tape X'd on windows across the street, against wind.

"And you'd see more police cars. Crimes go up in summers, with the heat.

"Do you think this sounds like science fiction? In a way it *is* science fiction," the scientist would tell the reporter. "But my wife and I used to go to the beach when we were first married. I'd look at the waves. Nature seemed so powerful, you have to wonder. These gasses people are sending into the atmosphere. Can they *really* compete with the powerful forces of nature? Even as a scientist you wonder. But after being involved in these studies over a decade, checking the models, looking at the Earth's climate, figuring out what kind of force is necessary to really change nature," he would say, as if he wished the truth was otherwise, "you come to the conclusion that yeah, man can change it."[15]

---

[15] *The Coming Storm: Extreme Weather and our Terrifying Future*, Bob Reiss, Hyperion, New York, 2001

To be fair, Hansen addresses criticism to this prediction—though with talking points, not studious, verifiable data.[16] Nonetheless, it's fair to test his prediction, isn't it? Let's grab a satellite shot off the web and see how this forecast is holding up...

Oops!

Figure 1

GISS Office in Manhattan[17]

Perhaps by the time you're reading this book, the situation will be

---

different, but twenty-four years after prognosticator Hansen's dire forecast, you don't need wading boots to stroll along the Henry Hudson Parkway.

As an aside, do you know what the sea level was 20,000 years ago? Answer: 140 meters lower than it is now (see Figure 2). Do you think human activity caused the massive sea level increase as we exited the last ice age? Believe me, when the inevitable new ice age begins, your heirs will have more important things than a few feet of sea level to worry about.

> *Who would have thought that as recently as 70,000 years ago, extremes of climate had reduced our population to such small numbers that we were on the very edge of extinction.* [18]
> —Meave Leakey, Paleontologist, Genographic Advisory Board member, National Geographic Explorer in Residence and Research Professor at Stony Brook University.

One thing I can say with reasonable certainty: if something happened in history, it will happen again. What does that tell you about your mental health if you're obsessing over a few inches of sea level rise? Are you bullish or bearish on owning waterfront property?

---

[18] http://phys.org/news128261865.html

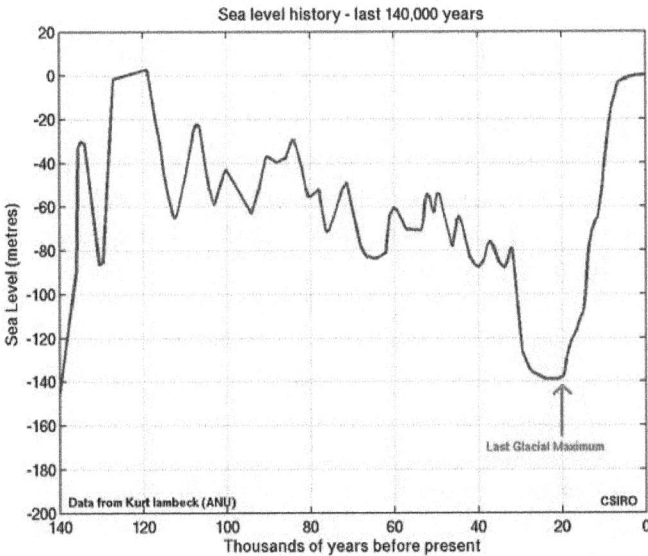

Figure 2

Historic Sea Level Changes[19]

I hate to mess with someone's livelihood—so I hope the devil will forgive me for writing the following letter to Hansen's former boss.

Dr. Michael Griffin June 24, 2008
NASA Headquarters
Suite 5K39
Washington, DC 20546-0001

Dear Dr. Griffin.
I am writing today to urge you to terminate the services of Dr. James E. Hansen. Dr. Hansen is not simply an embarrassment and blemish on the reputation of the National Aeronautics

---

[19] http://www.cmar.csiro.au/sealevel/sl_hist_intro.html#fewthousand

and Space Administration; he is responsible for great harm to the credibility of science. Worse than that, as the world cools (due mainly to reduced solar activity), we will find that his absurd, repeated warnings about human-caused global warming leads many to complacency when a more proper course of action would be preparing for cooling and all it carries with it, i.e. compressed food-producing latitudes and shorter growing seasons.

I have no idea what benefit the public gets from the employment of Dr. Hansen, but I doubt that "projecting humans' potential impacts on climate" is part of his legitimate job description or a role that justifies the consumption of tax dollars.

Again, please terminate the services of Dr. Hansen. The soapbox he enjoys at tax-payer's expense must be removed.

Thank you very much for considering my plea.

—Best regards;

Does this seem harsh? In general, I don't want people to be fired, but James 'Death Train' Hansen[20] is a raving lunatic. The idea that he gets a government paycheck (and worse, has a government expense account) really offends me.

Michael Griffin stirred up controversy (which he later apologized for) by making critical comments about human-cause global warming, comments Hansen called *ignorant*. Do I need to say it? I think Griffin was dead-on correct. It would be a happy day if Hansen ever called me ignorant (or a buffoon).

I have no doubt that global—that a trend of global warming exists. I am not sure that it is fair to say that it is a

---

[20] *The trains carrying coal to power plants are death trains. Coal-fired power plants are factories of death.*
—Dr. James E. Hansen,
http://www.guardian.co.uk/commentisfree/2009/feb/15/james-hansen-power-plants-coal

*problem we must wrestle with. To assume that it is a problem is
to assume that the state of Earth's climate today is the optimal
climate, the best climate that we could have or ever have had
and that we need to take steps to make sure that it doesn't
change.*

*First of all, I don't think it's within the power of human
beings to assure that the climate does not change, as millions
of years of history have shown, and second of all, I guess I
would ask which human beings—where and when—are to be
accorded the privilege of deciding that this particular climate
that we have right here today, right now is the best climate for
all other human beings. I think that's a rather arrogant
position for people to take.*[21]

—Michael Griffin

Interacting with the Author

It's easy to contact me—ken@StairwayPress.com. Fair warning:
this address attracts a lot of spam eaten by automatic filters. In
addition, I am often in a hurry when studying subject lines—and
quick on the delete button if I think someone is wasting my time.
So, if you have something interesting to say, please be persistent
and use a descriptive subject line.

Alternately, there is a snail mail address in the front matter of
this book. If all else fails, send me a note. I'm no hermit. I can be
reached.

I like talking about climate and other technical subjects. If you
find an error in my thinking, I'm pleased to hear about it.
However, if your commentary consists of insults or other dumb
stuff I've heard a million times before, then please don't bother.
I'm well aware the eminent folks at RealClimate think I'm a
buffoon—it's unnecessary to remind me.

---

[21]

http://www.spacedaily.com/reports/NASA_Administrator_Michael_G
riffin_Not_Sure_Global_Warming_A_Problem_999.html

THE GENIUSES OF REALCLIMATE

Owing to pressures on my time, I will not be able to respond to any further inquiries from you. Given your extremely poor past record of reporting on climate change issues, however, I will leave you with some final words. Professional journalists I am used to dealing with do not rely upon un-peer-reviewed claims off internet sites for their sources of information. They rely instead on peer-reviewed scientific research, and mainstream, rather than fringe, scientific opinion.
—Michael E. Mann, Director of Penn State University's Earth System Science Center

IN CASE YOU don't know, the man shown in Josh's mug shot (below) is Michael Mann[22], a cofounder of RealClimate.org (affectionately known as RC, for short).

---

[22] Dr. Michael E. Mann, Professor of Meteorology, Pennsylvania State University

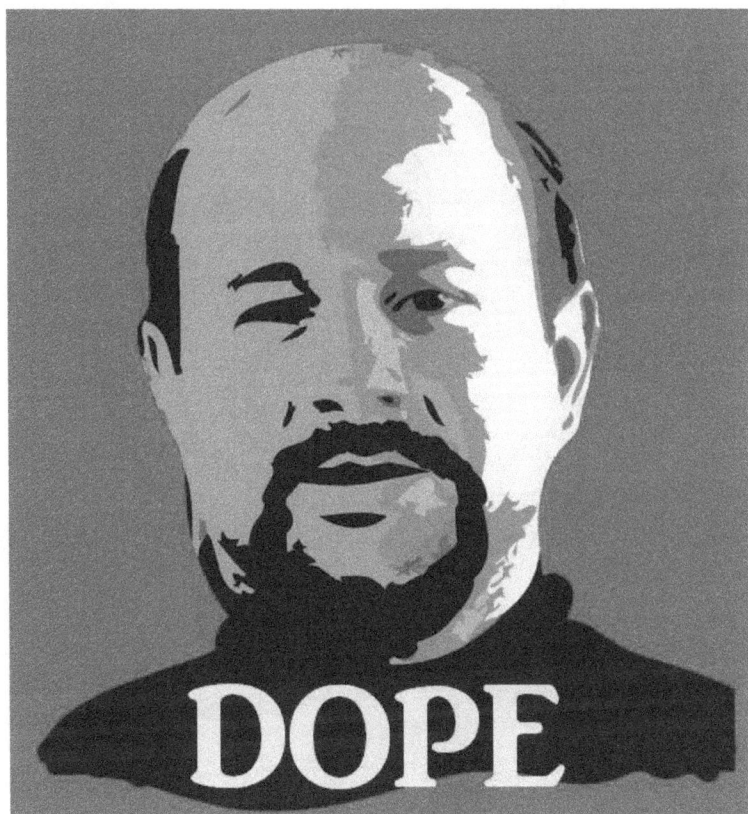

Figure 3

Dr. Michael Mann, an Example of a Modern Fool

The good folks at RC think I'm a buffoon? I'm a man, so, according to my wife, that assertion will always be at least partially true. Does this trouble me? Not even vaguely—I wear it as a badge of honor. An insult from an evil fool is a compliment.

Here's the real buffoon…

Figure 4

What can I say about RC? It promotes itself as a scientific site, but it's really an advocacy website and a clearinghouse for human-caused warming propaganda. They routinely censor contrary opinions and inconvenient criticism. I am a to-the-core believer in freedom of expression, so I don't object to their vociferous activism, but you can't trust them or their proclamations—they are liars and misleaders. Hell, just do a WHOIS search and look at who owns their web domain...

Domain ID:D105219760-LROR
Domain Name:REALCLIMATE.ORG
Created On:19-Nov-2004 16:39:03 UTC
Registrant Name:Betsy Ensley
Registrant Organization:Environmental Media Services
Registrant Street1:1320 18th St, NW
Registrant Street2:5th Floor
Registrant City:Washington
Registrant State/Province:DC
Registrant Postal Code:20036
Registrant Country:US
Registrant Email:betsy@ems.org

From a mini-bio I found online, Betsy Ensley "previously managed advocacy websites for Environmental Media Services and MoveOn.org in the build up to the 2004 election cycle."[23] Note the use of the word 'advocacy'.

Who is EMS? According to Wikipedia[24]:

> EMS was founded in 1994 by Arlie Schardt, a former journalist, former communications director for Al Gore's 2000 Presidential campaign, and former head of

---

[23] http://chicago2011.drupal.org/user/betsy-ensley
[24] Wikipedia can be trusted for some topics, but not for controversial topics leftards love—like global warming.

the Environmental Defense Fund during the 1970s.[25]

These parasites have a vested interest in keeping the incredible stream of easy cash flowing.

Here's a fair question: why am I so obsessed with this topic? It's not just the trillions of dollars the Warmistas want to waste on stupid stuff like exploding windmills and acres of ugly solar panels. It's not just the activist's desire to tear down western civilization—a civilization that has done so much to make modern life so comfortable for the average person with sanitation and medicine and cheap nutrition and summer cooling and winter heating and delivering us from boredom with online media and endless blogs to read and the ability to argue with knuckleheads from every part of the globe. Noisy activism makes me angry, but it does not feed the pool of red hot molten iron that burns in my belly.

In fact, this topic even pops up around the dinner table. The following question comes from Babe, the buffoon's daughter.

Why Do You Care So Much?

What a great question. I spend a lot of time studying the news of the day in climate science and dedicated significant money and energy to publishing *Slaying the Sky Dragon: Death of the Green House Gas Theory* and other books. I don't say much about the fine folks comprising the Sky Dragon Slayers in this book—most of that interaction and commentary will appear in Volume Two.

I believe there are fundamental types of people. One prominent group wants to run the world—for the overall benefit of the unwashed masses, of course. Call these people Collectivists. Progressive do-gooders. Communists. Watermelons. Meddlers. Petty tyrants. Bureaucrats. Nanny state busybodies. I hate them with an unhealthy passion. I do. I despise them. I do not want to be their slave and I do not want my children enslaved by them. This is

---

[25] http://en.wikipedia.org/wiki/Environmental_Media_Services

a central battle of my life and it's worth fighting.

It probably sounds paranoid and unfashionable—deserving of derision, but I readily admit to being influenced by Ayn Rand's objectivist epistemology.

She was an impatient, abrupt person and challenged people with questions like: "What is your basis?" In other words, what is your central philosophy of life? This is a tough question and perhaps unfair to the average person. But, you only have one life to live and you *should* know your basis. If you don't, you're living an unconsidered life.[26] Science and technology are central tenets of my religion. I take them seriously. They are my basis. The only way we can truly *know* anything is via the scientific method.

I believe the unfettered human spirit creates beautiful things like sewage treatment plants, progressive rock epics, surrealistic paintings and nuclear power plants. In fact, I believe *all* the notable achievements of our species came from the unleashed human spirit. Do you enjoy warmth in the winter and air conditioning in the summer? Are you breaking your back in a field? Are you starving? Can you cheaply travel nearly any place in the world? Do you have smallpox scars on your face? Are you coughing your guts out while slowly dying of Tuberculosis? If you like your answers to these questions, then you are enjoying the fruits of freedom and liberty. That's what I believe.

In addition, I *don't* believe people must be coerced to do good. In general, they will do the right thing on their own. Most of the human-caused evil in the world is caused by dictators, religious fundamentalists, totalitarian regimes and oppressive governments. I don't think it's an accident that rich, western—freer—countries are cleaner, safer countries. Many of the world's people want to escape third-world poverty and despair for a chance to be

---

[26] Acting through the fantasy of freedom, in existential rebellion, is not only the source of our dignity, moral choice, and responsibility, but also the ineluctable fundament of our values, lest we be nothing more than passive playthings of an indifferent universe or capricious gods.
From *What Matters: Most: Living a More Considered Life* by James Hollis.

productive and make a better place for their children—in the West.

There are many fronts in the battle. I am a person inspired by technology and science; and the battle I engage is with progressive activists who want a global system of taxes and regulation to save the world. They love cloaking their collectivist aspirations in science. Human-generated CO_2 will kill the world! What a fine theme song for our march to the worker's paradise—as we're herded along the sad road to serfdom.

Unfortunately for the progressive activists, their so-called *science* is nonsense. What's funny is it doesn't take much critical thinking to debunk their every tenet; even a buffoon like me can figure it out. Every assertion is not just wrong, but laughable. You have to give them credit for getting so far on such skeletal strands of data and hand-waving analysis—creating, literally out of cold, thin air—melodramatic emergencies to move their ambitions forward. Here is a partial list of climate nonsense:

- *The concentration of atmospheric CO_2 is correlated with surface temperatures.*
- *Current temperatures are unprecedented.*
- *Increasing temperatures are a hazard to our health.*
- *The current rate of temperature change is unique and alarming.*
- *There is evidence that human activities affect the global climate.*
- *The science is settled and there is overwhelming consensus among scientists.*
- *CO_2 is a forcing and the water vapor response is a feedback.*
- *Human-generated CO_2 accumulates in the atmosphere and creates an imbalance in the source/sink equilibrium.*
- *Greenhouse gases—stimulated by out-going infrared radiation—increase the Earth's average surface*

temperature by 33°C.

The list goes on and on—I'm sure you can think of some I missed.

So I care and I fight. And we're winning! Forgive me for rejoicing while the consensus falls apart and the increasingly desperate progressive activists wriggle and squirm. The world was so close to being theirs, but now they get to experience a massive, well-deserved *FAIL*.

That does not mean the war is over. Like all cynical, disgusting exploiters, they will move on to the next scheme—heaven forbid they get real jobs and do something productive like weaving blankets or working on oil rigs in Texas. Real work is not a suitable pastime for these elite thinkers. But, wielding the bright, sanitizing light of truth and scrutiny we're ready for the creeps where ever they arise.

So, can I ask?

What is your basis?

This book documents my personal and intensely subjective interaction with the community of climate activists. I think you'll find it amusing whether you agree with me or not. As an example of this interaction, I was once accused of not caring about leaving a habitable world for my children and grandchildren. This particularly moronic assertion inspired the following conversation with my beloved daughter (Babe).

KLC: Hey, Babe. I take a lot of heat from Lilliputian morons like DickWad. As a reality check, I'd like to hit you with a few questions if you're willing.

Would you say, from your point of view, that I don't care about your future and our following generations?

Babe: You've spent quite a bit of time over the years steering me and my offspring toward becoming educated and contributing to society, so no, I would not say that.

KLC: Unless I'm completely delusional, it seems like I have always encouraged, no, that's not the right word, I've *forced* you to think for yourself. Have I ever told you what to think about any topic?

Babe: Well, there was the Pig-Head Photo[27] incident when I was five, which was pretty leading. Aside from that, you've always encouraged me to think for myself, even though you disagree with most of what I come up with. :) It's not your fault; you're only a buffoon, right?

KLC: Let's say I was a progressive activist busybody—and I convinced you to drive a Smart car or a Prius or some other small, green politically-correct car. Would you be alive today?

Babe: Nope, I'd be roadkill[28]. I still want one though, so feel free to provide me with one and I'll do my very best to avoid being run over. Red would be great—maybe with a racing stripe?

KLC: I think it's safe to say you know me, for better or worse, much more than the delusional DickWad. Do you have any comments, suggestions or added information you'd like to address to him?

Babe: *Moron? Delusional DickWad?* That's not very nice. *I* was raised to respect the opinions of others—even when they are not in line with my own views. You boys should play nice. Perhaps agree to disagree? I know nothing about this person, but I would tell him you undoubtedly enjoy getting his goat and thrive on his negative feedback. I think it's great to have people examining issues from every angle, but if you both assume you're absolutely right and the other is absolutely wrong and neither of you want to have a civilized, open-minded, debate, then you're probably *both* buffoons! Move on! :)

I will say that I think you're a pretty smart guy and I appreciate that if you don't know the answer to a question, you are

---

[27] In this odd incident, a pretty picture brought home from the local Sunday School was molested: Jesus' handsome head was replaced with a pig's. Actually, that's still damned funny if you ask me.

[28] This comment refers to a serious car accident when an elderly, befuddled driver turned his car in front of her and she couldn't avoid hitting him. Her demolished Pathfinder ended up on its side. If she was driving a Chevy Volt, she'd be dead and I'd be bereft.

comfortable saying so. I'm confident that you have very strong evidence to back up any side you're willing to commit yourself to. Now if you'll excuse me, I need to take out my recycle bin and possibly hug a tree while I'm out there.

Science is My Religion

There is a bit more going on. As I said, science and technology are my avocation and religion. I've worked on flight-critical avionics systems where there is no luxury of assuming the physical universe will conform to the way I want it to work. In engineering, there are consequences to getting things wrong—ideas must be tested and validated. I take this seriously and it offends me when my religion is trampled on by the muddy boots of catastrophic alarmists.

Something you'll notice about Warmista activists is their projection. Their propaganda efforts are well-funded by governmental and non-governmental agencies. They assume their enemies do what they do. I defy you to find billions of dollars spent promoting skeptical climate science. At the same time, these huge sums are easy to find in funding by governmental agencies (like NASA and the EPA) and non-governmental agencies (like Greenpeace, The Sierra Club, Natural Resources Defense Council and The World Wildlife Fund).

As mentioned earlier, one of the kingpins of global warming hysteria is Michael Mann. For a complete deconstruction of his moronic hockey stick graph, see A. W. Montford's excellent book.[29]

> The evidence for a well-organized, well-funded, and orchestrated climate change disinformation campaign has been laid out in detail on public interest group Web sites, in articles in popular magazines, and by an increasingly rich

---

[29] *The Hockey Stick Illusion: Climategate and the Corruption of Science*, A. W. Montford, Stacey International, 2010.

array of scrupulously researched books on the topic.[30]
I think Pfizer is missing a great promotional opportunity if they don't contract Michael Mann for using his face on promotional posters for Xanax. Is this guy paranoid, or what?

For the record, I don't hate all the proponents of human-caused global warming. I disagree with them, often vehemently, but there are some who appear to be completely decent people. As examples, I have exchanged emails with Spencer Weart[31], Judith Curry[32], Grant Petty[33] and others. For this subset, I find them bright, patient and polite.

If all the warriors on their side were such, then the battle lines would not be so hardened and much of the vitriol and gnashing of teeth could be avoided. Besides, it should not matter much what I think or what Gavin Schmidt[34] thinks. Our differences are settled at the ballot box when the general public chooses a candidate that represents their views the most and we move on with our lives.

However, the Warmistas don't think the general public is bright enough to make the proper rational decisions. They want to drive policy with dictates from bureaucratic totalitarian organizations like the United Nations. One can only imagine the green utopia of which they dream. Of course, it would not be very green because it will be littered with shattered pieces of wind

---

[30] *The Hockey Stick and the Climate Wars*, Michael E. Mann, Columbia University Press, 2012, p. 63.

[31] Former director of the Center for History of Physics of the American Institute of Physics and author of books including *The Discovery of Global Warming*, Harvard University Press, 2003 (updated: 2008).

[32] Climatologist and chair of the School of Earth and Atmospheric Sciences at the Georgia Institute of Technology.

[33] Professor of Atmospheric Science, University of Wisconsin-Madison and author of *A First Course in Atmospheric Radiation* and *A First Course in Atmospheric Thermodynamics*, both published by Sundog Publishing. Source: Grant Petty website: http://sleet.aos.wisc.edu/~gpetty/wp/

[34] Climatologist and climate modeler at the NASA Goddard Institute for Space Studies. He's one of the cofounders of RealClimate.org

turbines, mutilated bird parts, acres of broken, birdshit-stained solar panels and forests you won't be allowed to visit for fear of disturbing a spotted owl.

Perhaps this is as good a place any for a full disclosure. I am not funded by any outside organization. This includes The Heartland Institute and the Koch brothers.

In fact, my situation is the opposite; I pay my bills by selling books to the public and from a generous salary generated by my day job as an electrical engineer. The company I currently work for[35] is very excited about alternative energy and spends significant research dollars on *green* power conversion solutions. If they were to state an opinion, they would be displeased at my thoughts and how freely I share them. Hell, I can prove it. Just look at the slide below which was presented at a tradeshow[36] by our Chief Technology Officer, Dan Kinzer.

Why do I hate Michael Mann so much? His dim-witted hockey stick graph is ubiquitous. I know some people that work for my company fully embrace the causes of *green* technology while others think along the same lines as me.

I would love to get a grant from the Koch brothers—I'd be proud to take it, but so far, the big money has not been forthcoming. If you ever see me driving a Maserati or Maybach, it might be safe to assume this situation has changed. Ask me and I'll tell you straight up.

---

[35] Fairchild Semiconductor, NYSE: FCS
[36] APEC (Applied Power Electronics Conference and Exposition) 2012, http://www.apec-conf.org/images/PDF/2012/Plenary/4-2012_plenary_presentation-kinzer.pdf

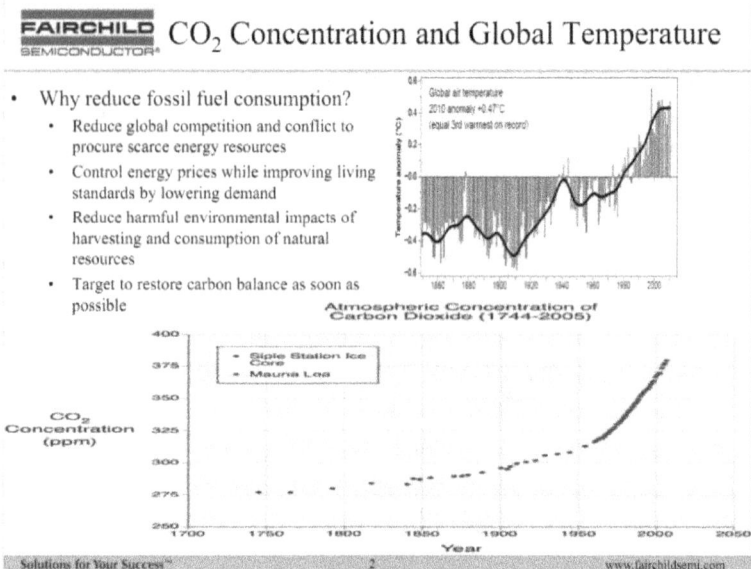

FAIRCHILD CO_2 Concentration and Global Temperature
SEMICONDUCTOR

Figure 5

A Random Example of the use of Michael Mann's Stupid Hockey Stick

Forms of Effective Argumentation

A lot of the people who argue about climate science and public policy are not stupid; they are very effective propagandists. The alert skeptic will notice the way an argument is manipulated. Here is a list of things they do. I'm not saying this is necessarily wrong. Hell, to be effective, maybe we need to do the same thing right back at them. There are lots of ways to frame the facts to drive home a central point. Regardless, we should watch for their logical fallacies and call them on it.

- **Argumentum Ad Hominem** This is Latin which roughly means attacking the man. You see examples of this all the time and it's an effective method of diverting attention from the argument at hand. An example is to say noted skeptic Fred Singer is wrong because he once took money from an oil or tobacco company. What does that have to do with the facts of the argument? That's right, nothing. Good call.

- **Appeal to Authority** Again, the point here is to shut down debate by referring to the credentials of someone who has a point of view. There are plenty of people with PhDs who are whack-a-doodle crazy. I disagree with Dr. Gavin Schmidt about a lot of things. He has a PhD and I don't. Is that a sufficient condition to validate his argument? If you think so, then this is the wrong book for you. Ask for your money back immediately. Instead, send your money to Gavin's early retirement fund.

- **Tautological Argument** This is a circular argument where point A is used to support point B and point B is used to support point C. However, if the only support for A is C, then the argument is circular and has no merit.

- **Confusing Correlation with Causation** Correlations are cheap and common, whereas uncovering root causes can be devilishly difficult. For example, pointing out temperatures increased at the same time atmospheric CO_2 increased proves nothing. You have to isolate, identify and repeatably demonstrate the cause-effect linkages between the two.

- **False Dichotomy** In this argument, it is falsely assumed that a situation is either one way or another, when in the real situation, there are many more possibilities. How do you rigorously exclude all the unknown unknowns?

- **Red Herring.** Bringing up irrelevant points to distract from the argument you're losing.

- **Straw Man Argument** This is where you falsely state an opponent's position, then tear down the bogus position.

- **Texas Sharpshooter Fallacy** In this case, data is

collected to support an argument. The proper way to support an argument is to make a prediction, then collect the data to test to see if the argument was supported. A Texas Sharpshooter shoots the target, then walks up and draws the bull's-eye around the bullet holes.

- **Godwin's Rule of Nazi Analogies** Godwin observed that, given enough time, in *any* online discussion—regardless of topic or scope—someone inevitably criticizes some point made in the discussion by comparing it to beliefs held by Hitler and the Nazis.[37]

- **Noble Cause Corruption** In this aspect of human nature, the end cause is so important and righteous that fudging the facts and acting dishonestly is justified in the mind of the activist. The end justifies the means, right?[38] Don't quite have all the facts to support the storyline you're selling? No problem, just make stuff up.

- **Confirmation Bias** Humans are pattern-seeking, story-telling animals. This means we need to be very careful about random, chaotic data because we have a natural instinct to project meaning into it.

- **Projection** One thing you'll notice about the progressive activists: they play dirty and assume you are too. They exercise all kinds of sleazy tactics and automatically assume their opponents are doing the same. Watch what they accuse you of, because that's what they are doing.

- **Selective Error Correction** Suppose you're looking at a scatter plot and some data doesn't support your theory—

---

[37] Source: http://en.wikipedia.org/wiki/Godwin%27s_law
[38] That perennial question, "Does the end justify the means?" is meaningless as it stands; the real and only question regarding the ethics of means and ends is, and always has been, "Does this *particular* end justify this *particular* means?"
—*Rules for Radicals*, Saul Alinsky.

you could study the data and throw it out due to discovered errors. However, to be an honest researcher, you need to apply the same rigor to the data that supports your theory.

Frickin' Mathmeticians

Were academic mathematicians on my list of people I don't like? They should be. Here's why their work should be viewed with suspicion. Let's look at a simple formula:

$$a = b/10$$

That looks innocent enough. Let's convert it to an equivalent formula by multiplying both sides by 10—which is perfectly legitimate mathematics, right? As long as you so the same thing to both sides, mathematically, we're okay.

$$b = 10a$$

So, what is the problem? Imagine the first equation was based on a cause-effect observation where we collected data and noticed that when we wiggled 'b', 10% of this stimulus was seen at 'a'. Changing 'b' *caused* 'a' to change. However, is there any physical justification for the equivalent version shown in the second equation? In the second version of the same damned equation, 'b' is shown to be very sensitive to small changes of 'a'.

Not all processes are reversible. To pick a random example, if the lousy bastards at the IRS take 20% of my income, that does not mean an increase in IRS revenue reflects backwards into my bank account. In fact, the real result is probably the opposite: they'll use the added revenue to hire more enforcement agents to oppress me even more.

So, I suggest caution when academic climatologists play with their formulas. This advice also holds for computer models which are, after all, just very complex formulas.

Don't be Fooled by Complexity

There is a wise maxim:

> *When you can't dazzle them with brilliance, baffle them with bull.*[39]

Getting into the details of anything can lead to incredible difficulty and complexity. However, just because something is complex does not mean it's correct. In fact, you can lead a good, safe life if you assume exactly the opposite. Regardless of the complexity, there should be a summary that is easy to understand and anyone that can't provide that? As my grandfather said, it's time to hold onto your wallet.

The Cupcake Criteria

Let's apply the Cupcake Criteria to the problem of global warming. Any parent can relate to this. Suppose you have a cupcake to be divided between two ravenous children. How do we avoid bloodshed and make sure the kids are treated fairly? Easy! Have one kid cut the cupcake and have the other select which half to consume. I assure you, the cupcake will be cut very carefully and the halves will be as equal as is humanly possible.

Why do I bring up something so simpleminded? Because you should be on high-alert when dealing with situations where one vested interest cuts the cake and also selects the beneficiaries.

Here's how this relates to climate science: when government funding is used for academic projects (directly and through NGOs), how is the taxpayer involved to make sure the money is spent wisely? For organizations funded from private sources, who cares? They should do what they want, but much of the money spent on environmentalism is extracted from the taxpayer.

---

[39] W. C. Fields

Ken Coffman

In these papers, notice how they ask for additional projects (and additional funding, of course).

> Further work is required to establish relationships, if any, between the clear-sky and all-sky radiative cooling and precipitation rate in the present climate on interannual and decadal to longer timescales.[40]

> The suggested method can be applied to other dataset of highly-resolved infrared spectra (e.g. to measurements made within the Network for Detection of Stratospheric Change).
> However, the capability of the method would have to be investigated for each measurement site individually. [...]
> Applying the proposed retrieval method to spectra measured during the last 20-25 years at FTIR sites such as Jungfraujoch or Kitt Peak could produce unique continuous long-term series of UT/LS water vapour amounts.[41]

> So, how can future research improve upon earlier studies (Stahle et al. 1998; Mann et al. 2000; Evans et al. 2002. D'Arrigo et al. 2005a)? Ultimately, the central and eastern Pacific are the key locations where new annually resolved proxy records need to be developed.[42]

Pathetic, isn't it? Pick any modern government-funded paper and part of the author's requisite goal is keeping the gravy train going.

---

[40] *Variability in clear-sky longwave radiative cooling of the atmosphere*, Richard P. Allan, Journal of Geophysical Research, VOL. 111, 2006.

[41] *Water vapour profiles by ground-based FTIR spectroscopy: study for an optimised retrieval and its validation*, M. Schneider, F. Hase, and T. Blumenstock, Atmosheric Chemistry and Physics, Vol 6, 811-830, 2006

[42] Reconstructing ENSO—Methods, Proxy Data and Teleconnections, Rob Wilson, Edward Cook, Rosanne D'Arrigo, Nadja Riedwyl, Mike Evans, Alexander Tudhope and Rob Allan. Journal of Quaternary Science, 2010 25(1) 62-7.

Hell, without grants, these folks might have to do something useful like being a grease monkey at a quick-change oil joint.

Let's look at an example of government funding. The National Science Foundation (NSF) has a Dimensions of Diversity.[43] From the synopsis of their mission:

> *Despite centuries of discovery, most of our planet's biodiversity remains unknown. The scale of the unknown diversity on Earth is especially troubling given the rapid and permanent loss of biodiversity across the globe. With this loss, humanity is losing links in the web of life that provide ecosystem services, forfeiting an understanding of the history and future of the living world, and losing opportunities for future beneficial discoveries in the domains of food, fiber, fuel, pharmaceuticals, and bio-inspired innovation.*

If you click around, you will find they funded programs which add up to $45,815,656. The NSF is requesting a 2013 budget of $7,373,100,000.[44] Let there be no doubt, there is a lot of money flying around—which funds lots of projects. These numbers come from a NSF Infobrief: "With Help from ARRA[45], Universities Report $61 Billion in FY 2010 Total R&D".[46]

Here's fair warning for rent-seeking academics: opulent government funding is a bubble. You don't need to be an expert in Young-Laplace Equations[47] to know what always happens to

---

[43]

http://www.nsf.gov/funding/pgm_summ.jsp?pims_id=503446&org=NSF&sel_org=NSF&from=fund

[44] http://www.nsf.gov/pubs/2012/nsf12045/nsf12045.pdf

[45] American Recovery and Reinvestment Act. This is Barack Obama's $787,000,000,000 "stimulus" package. You can call it a slush fund if you like.

[46] http://www.nsf.gov/statistics/infbrief/nsf12313/nsf12313.pdf

[47] In physics, the Young-Laplace equation is a nonlinear partial differential equation that describes the capillary pressure difference sustained across

bubbles. Remember, friends:

Things that can't go on forever, don't.[48]

Why?

Why did I write this book? While I have often been furiously angry at the exchanges documented in this book, mostly it's been fun and I've learned a lot. As the global warming movement falls apart, I think it will be amusing to have a humorous memoir about the heady days when the battle was fully engaged. By my estimate, the activists were winning until about 2007; they had profligate funding and the mainstream media mined their papers for pithy, voice-of-doom headlines. Life was good for them, but then the tide turned and now they are on the run.

Let's not forget how stupid, en masse, our species can be. Let's be ready for the next round of nonsense from do-gooder activists who will switch from global warming to some other form of bullshit like species eradication, increased climate extremes, human-caused hurricanes, habitat destruction, ocean acidification or what? These people are creative in their madness, so who knows what they'll come up with next—let's learn from the past and prepare for the next battle.

---

the interface between two static fluids, such as water and air, due to the phenomenon of surface tension or wall tension, although usage on the latter is only applicable if assuming that the wall is very thin. Source: Wikipedia

[48] Herbert Stein, chairman of the Council of Economic Advisers during the Nixon administration

THE GIANTS OF HISTORY

The fact is that we can't account for the lack of warming at the moment and it is a travesty that we can't.[49]
—Kevin E. Trenberth, head of the Climate Analysis Section of the U.S. National Center for Atmospheric Research (NCAR)

IF YOU HANG out with global warming activists, they'll often bring up the established physics of yore from brilliant pioneers like Joseph Fourier, John Tyndall and Svante Arrhenius. This is what these folks mean when they make stupid comments like the GHG effect on climate was discovered 150 years ago.

> *As a basis for discussion about GHGs and their influence on the climate, it should be noted that there is a natural, non-*

---

[49] In fairness, he later said this: *It is amazing to see this particular quote lambasted so often. It stems from a paper I published this year bemoaning our inability to effectively monitor the energy flows associated with short-term climate variability. It is quite clear from the paper that I was not questioning the link between anthropogenic greenhouse gas emissions and warming, or even suggesting that recent temperatures are unusual in the context of short-term natural variability.*
Heaven forbid his credentials as a corrupt progressive activist be questioned.

> anthropogenic greenhouse effect, which Joseph Fourier
> discovered more than 150 years ago.
>
>   Fourier argued that "the atmosphere acts like the glass of
> a hothouse because it lets through the light rays of the Sun but
> retains the dark rays from the ground".[50] This is a major
> simplification in describing the greenhouse effect, but it does
> provide insight into why the Earth's surface is considerably
> warmer than it would be without an atmosphere.[51]

That's not what Fourier said[52], but never mind that now. Academic climatologists throw out the names Tyndall and Arrhenius like they are magic words—as if these names nullify all opposing arguments. Of course, these legendary heroes are not around to defend themselves or improve their thinking by taking into account the results of added information from much improved instrumentation, so it falls to buffoons like me to do so. I admire these people very

---

[50] Joseph Fourier, *Remarques Générales Sur Les Températures Du Globe Terrestre Et Des Espaces Planétaires*, 27 ANNALES DE CHIMIE ET DE PHYSIQUE p.136–67 (1824) *and* Joseph Fourier, *Mémoire Sur Les Températures Du Globe Terrestre Et Des Espaces Planétaires*, 7 MÉMOIRES DE L'ACADÉMIE ROYALE DES SCIENCES p.569-604 (1827).

[51] *A Rational Discussion of Climate Change: the Science, the Evidence, the Response,*

http://gop.science.house.gov/Media/hearings/energy10/nov17/charter.pdf

[52] *In short, if all the strata of air of which the atmosphere is formed, preserved their density with their transparency, and lost only the mobility which is peculiar to them, this mass of air, thus become solid, on being exposed to the rays of the sun, would produce an effect the same in kind with that we have just described. The heat, coming in the state of light to the solid earth, would lose all at once, and almost entirely, its power of passing through transparent solids: it would accumulate in the lower strata of the atmosphere, which would thus acquire very high temperatures. We should observe at the same time a diminution of the degree of acquired heat, as we go from the surface of the earth.*

—Jean Baptiste Joseph Fourier, *General Remarks on the Temperature of the Terrestrial Globe and the Planetary Spaces* much and it offends me when their holy names are used in vain.

I've read a lot of Tyndall's work and he was a genius, there's no doubt about that.

> Experiments are next described which illustrate the action of the aqueous vapour of our atmosphere on radiant heat; and considerations follow regarding the influence of an atmosphere like ours upon the temperature of a planet. In former speculations upon this subject the density and height of the atmosphere were dwelt upon by distinguished writers; but it is here pointed out that a comparatively slight change in the variable constituents of our atmosphere, by permitting free access of solar heat to the Earth, and checking the outflow of terrestrial heat towards space, would produce changes of climate as great as those which the discoveries of geology reveal.[53]

Here we have it, a great man like John Tyndall believed small proportions of atmospheric GHGs could change the surface temperature of the Earth.

> In 1859 Tyndall began to study the capacities of various gases to absorb or transmit radiant heat. He showed that the main atmospheric gases, nitrogen and oxygen, are almost transparent to radiant heat, whereas water vapour, carbon dioxide and ozone are such good absorbers that, even in small quantities, these gases absorb heat radiation much more strongly than the rest of the atmosphere.
>
> Tyndall concluded that water vapour is the strongest absorber of heat in the atmosphere and is the principal gas controlling surface air temperature by inhibiting leakage of

---

[53] *Contributions to Molecular Physics in the Domain of Radiant Heat*, p. 4, D. Appleton and Company, 1873.
http://archive.org/stream/contributionsto03tyndgoog

the Earth's heat back into outer space. He declared that, without water vapour, the Earth's surface would be 'held fast in the iron grip of frost'—the greenhouse effect.

The greenhouse effect works as follows. Most of the Sun's energy is radiated as visible light. This is not absorbed by the atmosphere and passes through to warm the Earth. The warm Earth radiates heat back into the atmosphere as infrared radiation. This is avidly absorbed by atmospheric water vapour and carbon dioxide, trapping the heat and preventing the Earth from freezing[54].

Trapping the heat. That's very interesting. The atmosphere has a thermal capacity, of course it does. It can store thermal energy. And everything with a mass and temperature will radiate, of course it does. The question is: how can a low-density gas change the temperature of something with a high density and correlated large thermal capacity—like the water that covers 71% of the Earth's surface?

Many misunderstand how green houses work. Over and over, you can crack open a physics textbook and see unsupported statements like this:

The solar energy reaching the surface of the Earth is concentrated in short wavelengths, which can easily penetrate the greenhouse gases, such as Carbon Dioxide and Methane. The Earth, however, is cooler than the Sun and it radiates its heat in the form of energy in the far infrared range. These longer wavelengths are partially absorbed by the greenhouse gases and some of the solar heat is returned to Earth. At a certain temperature these processes are in equilibrium and the surface temperature of the Earth is stable. However, if more greenhouse gases are put in the atmosphere the amount of trapped terrestrial radiation increases, leading to an increase

---

[54] http://www.ucc.ie/academic/undersci/pages/sci_johntyndall.htm

in global temperature.[55]

How can we dispel this mythical greenhouse theory? It turns out we don't need to—its inaccuracy was addressed long ago.

Professor Wood's Greenhouse Experiment

It used to be assumed that a greenhouse—made of glass or plastic that blocks IR—is heated due to IR retained in the structure. However, as proved by R. W. Wood in 1909 and verified in 2011 by Nasif Nahle[56], a greenhouse works by restricting convection. It is not a radiation trap, it's a heated-air trap. From 1909 on, let's accept no hand waving and lame-ass excuses.

Let's look at Professor Wood's letter:

> *There appears to be a widespread belief that the comparatively high temperature produced within a closed space covered with glass, and exposed to solar radiation, results from a transformation of wave-length, that is, that the heat waves from the Sun, which are able to penetrate the glass, fall upon the walls of the enclosure and raise its temperature: the heat energy is re-emitted by the walls in the form of much longer waves, which are unable to penetrate the glass, the greenhouse acting as a radiation trap.*
>
> *I have always felt some doubt as to whether this action played any very large part in the elevation of temperature. It appeared much more probable that the part played by the glass was the prevention of the escape of the warm air heated by the*

---

[55] *The People's Physics Book*, James H. Dann and James J. Dann.
http://scipp.ucsc.edu/outreach/PPBFullbook.pdf
[56] *Experiment on the Cause of Real Greenhouses' Effect—Repeatability of Prof. Robert W. Wood's experiment*, Nasif Sabag Nahle,
http://www.biocab.org/Wood_Experiment_Repeated.html
From the conclusion: *The experiment performed by Prof. Robert W. Wood in 1909 is absolutely valid and systematically repeatable.*

ground within the enclosure. If we open the doors of a greenhouse on a cold and windy day, the trapping of radiation appears to lose much of its efficacy. As a matter of fact I am of the opinion that a greenhouse made of a glass transparent to waves of every possible length would show a temperature nearly, if not quite, as high as that observed in a glass house. The transparent screen allows the solar radiation to warm the ground, and the ground in turn warms the air, but only the limited amount within the enclosure. In the "open," the ground is continually brought into contact with cold air by convection currents.

To test the matter I constructed two enclosures of dead black cardboard, one covered with a glass plate, the other with a plate of rock-salt of equal thickness. The bulb of a thermometer was inserted in each enclosure and the whole packed in cotton, with the exception of the transparent plates which were exposed. When exposed to sunlight the temperature rose gradually to 65°C, the enclosure covered with the salt plate keeping a little ahead of the other, owing to the fact that it transmitted the longer waves from the Sun, which were stopped by the glass. In order to eliminate this action the sunlight was first passed through a glass plate.

There was now scarcely a difference of one degree between the temperatures of the two enclosures. The maximum temperature reached was about 55°C. From what we know about the distribution of energy in the spectrum of the radiation emitted by a body at 55°, it is clear that the rock-salt plate is capable of transmitting practically all of it, while the glass plate stops it entirely. This shows us that the loss of temperature of the ground by radiation is very small in comparison to the loss by convection, in other words that we gain very little from the circumstance that the radiation is trapped.

Is it therefore necessary to pay attention to trapped radiation in deducing the temperature of a planet as affected

47

by its atmosphere? The solar rays penetrate the atmosphere, warm the ground which in turn warms the atmosphere by contact and by convection currents. The heat received is thus stored up in the atmosphere, remaining there on account of the very low radiating power of a gas. It seems to me very doubtful if the atmosphere is warmed to any great extent by absorbing the radiation from the ground, even under the most favourable conditions.

I do not pretend to have gone very deeply into the matter, and publish this note merely to draw attention to the fact that trapped radiation appears to play but a very small part in the actual cases with which we are familiar.[57]

That seems clear enough, doesn't it? A greenhouse works by restricting convection and trapping warm air, not by restricting longwave radiation. Want to prove it? On a hot day, close your car windows and let the interior hear up. Now open your windows just a little to allow air to circulate without changing the IR emission very much. If trapped IR is the secret to greenhouse heating, then the car temperature won't change much. Is that what you notice? I thinketh not. Please don't be a climate scientist and adapt your observations to your pet theory. If your theory does not conform to observation, then discard the theory—it's wrong.

To show what a fair soul I am, I will show you the most compelling argument I've seen from the other side. If there are other intelligible critiques of Professor Wood's thoughts I'd be happy to hear about them.

Charles Greeley Abbot wrote a note[58] ostensibly refuting

---

[57] Philosophical Magazine in 1909 (Vol. 17, pp. 319-320): XXIV. *Note on the Theory of the Greenhouse* By Professor R. W. Wood (Communicated by the Author)
Found at: http://sci.tech-archive.net/Archive/sci.physics/2008-04/msg00422.html
[58] *The London, Edinburgh and Dublin Philosophical Magazine and Journal of Science*, Series 6, Vol.18, pp. 32-35.

Wood. Here follows the Wood-debunking note in its entirety.

Note on the Theory of the Greenhouse

—*C. G. Abbot, Director, Astrophysical Observatory, Smithsonian Institution.*

In a paper of the above title[59] Professor R. W. Wood states that he has compared two "hot-boxes" one having a glass cover, the other a cover of rock salt, but otherwise similar. A glass plate was interposed in the path of the entering Sun rays. He observed a maximum temperature of about 55 °C within each box when exposed to the Sun. He concludes that the function of the cover is mainly to prevent the loss of heat by convection, rather than the escape of long wave rays, and asks: "Is it therefore necessary to pay much attention to trapped radiation in deducing the temperature of the planet as affected by its atmosphere?"

It may interest some to know that much higher temperatures can be reached within a "hot-box" than that observed by Professor Wood, if precautions are taken to diminish the loss of heat by convection from the warmed outer surface of the cover. On November 4, 1897, the thermometer recorded 118 °C within a circular wooden box 50 centimeters in diameter, 10 centimeters deep, insulated in feathers, covered with three superposed and separated sheets of plate glass and exposed normally to the Sun rays of the yard of the Astrophysical Observatory in Washington. The temperature outside was 16 °C.

Agreeing with Professor Wood that the main function of the cover of a "hot-box" or hot-house" is to prevent loss of heat by convection, it is interesting to see if this could be predicted. Published experiments on the cooling of solids in dry air and

---

[59] Phil Mag. 6th Series, vol. xvii. p.319 (1909).

in vacuum give the relative rates of loss by convection and radiation under known circumstances. Planck's radiation formula for the "black body" enables computations to be made of the losses by radiation for different temperatures of source and sink. This transmission of glass, salt, and the water vapour of the atmosphere, and the effective temperature of the latter are approximately known. I have attempted to compute from such data the relative hindrance which salt and glass covers would interpose to the loss of heat by convection and radiation combined from a "black" surface at $55\,^{\circ}C$. For the dependence of the temperature of the Earth's surface on the atmosphere, some numerical data can be assigned also, and is shown below there is reason to think that "trapping" is more important perhaps than Professor Wood thinks.

From an interesting paper of P. Compan[60] it may be seen that for a blackened copper ball 2 centimeters in diameter cooling from a temperature of $55\,^{\circ}C$ to nearly "black" surroundings at $0\,^{\circ}C$, the rate of loss of heat by convection in still dry air at atmospheric pressure is four-thirds as rapid as the simultaneous loss by radiation. In a breeze of 3 metres per second the convection loss becomes 3 times as rapid as in still air, or 4 times as rapid as the loss by radiation. The loss of heat by convection alone is approximately proportional to the difference of temperatures between the source and the sink.

If the covers had been absent in Professor Wood's experiments, the boxes would have been exchanging radiation principally with the water vapour of the lower atmosphere. Experiments of Langley, Rubens and Aschkinass, and others indicate[61] that less than 10 per cent of the radiation from the Earth's surface can penetrate the water vapour of the atmosphere above a coast station like Baltimore. Hence the

---

[60] Annuales de Chimie et de Physique, t. xxvi. Pp 488-574 (1902)

[61] See Annals, Astrophysical Observatory, Smithsonian Institution, vol. ii pp 167-172

water vapour of the atmosphere can be considered as practically a "black body" for waves of great wave-length. The effective temperature of the water vapour layers with which a coverless 'hot-box" would have been exchanging radiation may be estimated at 0°C. If glass is interposed the radiation is entirely cut off. If a rock-salt plate 1 centimetre thick is interposed between the body at 55°C and surroundings at 0°C, the absorption of the plate is about 19 per cent[62] and the reflexion probably nearly 10 per cent more, so that the transmission may be reckoned at about 70 per cent.

In combining the preceding results with those of Compan we will at first neglect the loss of heat by convection from the outside of the cover. We may assume the temperature of the air just outside the "hot-box" to be 15°C and also that Newton's law of cooling is applicable to the convection loss. In still air with no cover the rate of loss of heat towards the front by convection and radiation combined is proportional to:

$$\frac{40}{55} * 133 + 100 = 197$$

With glass cover: $0 + 100 * 0.00 = 0$
With salt cover: $0 + 100 * 0.70 = 70$

Thus of the heat which would have escaped toward the front from a coverless box at 55°C in still moist air at 15°C the salt hinders $\frac{197-70}{197} = 63$ per cent as much as the glass. Remembering, however, that owing to its higher absorbing power for long wave rays the glass will be warmed more than the salt, the convection loss from the outside of the warmed cover will be greater for glass than salt, so that the efficiency of a salt cover may be much more than 65 per cent of the

[62] See Kayser's Handbuch d. Spectroscopie, vol. iv p. 485, and Planck's formula of radiation.

glass. The convection loss from the front of the cover is a considerable factor, for in the "hot-box" tried at this observatory the front of the inner glass cover became too hot to handle and often cracked with the heat.

In view of these figures we may agree with Professor Wood that a salt cover[63] is nearly as efficient as a glass one for a "hot-box", although it would seem strange that he observed no difference at all. Perhaps in spite of the glass filter the cover-glass obstructed the entering Sun rays more than salt. But is not the case quite different with a planet?

Let us take the mean temperature of the Earth's surface at 14°C, the mean effective temperature of the water-vapour layers to which it principally radiates as 0°C, the temperature of space a -273°C. Then the rates of escape of heat from the surface by radiation, first with the water-vapour layer interposed, and, second, imagining the air to be completely transparent to Earth rays, would be in the ratio of 19 to 100 according to Planck's formula. It is very difficult to estimate how fast the heat of the Earth's surface escapes by convection, because neither the difference of temperature between the surface and the air nor the rate of motion of the air is well known. But if for the sake of discussion we suppose a temperature difference of 10°C and a velocity of 3 metres per second, the rate of convection loss comes out a 0.54 as great as the rate at which heat would escape by radiation if the air was no hindrance. This assumed convection loss is 2.8 times as great, on the other hand, as the estimated rate of escape of heat by radiation to the water-vapour layers at 0°C. In other words, according to this estimate the convection is the main agent in removing heat from the Earth's surface as things are, but would be only a small factor if the air was transparent to long-wave rays.

If these figures represent at all the order of magnitudes of

---

[63] A salt cover, however, is better than a perfectly transparent one.

these quantities there can be no doubt, I think, that the atmosphere is important as a trapping agent to increase the Earth's surface temperature.

A fair estimate of the actual increase of the Earth's temperature due to "trapping" has been made. Imagine a perfectly "black" and rapidly rotating planet of the Earth's dimensions, situated beyond the orbit of the Earth at such a distance from the Sun that the radiation absorbed by it would be equal to that available to be absorbed by the Earth, allowing for the reflexion of clouds, et cetera. Such a planet would assume the temperature of -17°C, whereas the real Earth has a mean temperature of +14°C. The difference, 31°C, is attributable to three causes:

The imperfect "blackness" of the Earth.
The "blanket effect" of the atmosphere.
The warming of the Earth's surface by radio-active substances and internal heat.

Remembering that the Earth is mainly water covered, it must be almost "perfectly black" for long-wave rays. I myself regard the conduction of internal heat and that of radio-active substances as negligible. This would leave the full 31° as due to the "blanket effect."

If there were no water on the Earth, the emissive power of its surface for long-wave rays would be less. Also, on account of the absence of clouds, its absorption of solar rays would be greater. The two differences would perhaps more than counterbalance the loss of the "blanketing effect," so that the mean temperature of the Earth without water would perhaps be rather higher than now, but much less uniform, ranging from above present temperatures by day to far below 0°C by night.

Now you can judge for yourself—who do you choose to believe?

Who do you put your confidence in? My colleague Nasif Sabag Nahle spent many hours checking to see if he agrees with Professor Wood. You can examine and critique the results of his experiment to your heart's content.[64] Here is his conclusion:

> *The greenhouse effect inside greenhouses is due to the blockage of convective heat transfer with the environment and it is not related, neither obeys, to any kind of "trapped" radiation. Therefore, the greenhouse effect does not exist as it is described in many didactic books and articles.*
>
> *The experiment performed by Prof. Robert W. Wood in 1909 is absolutely valid and systematically repeatable.[65]*

If you're the kind of person who will accept theory contrary to data collected by replicable experiment, then there is no hope for you. Enjoy your tenure in academia. I have useful work to do—stay out of my way. As an example of this kind of "thinking", I offer Joel D. Shore, PhD.[66]

Here is an exchange with Joel from WUWT.[67]

> *As an engineer, I am constantly seeking and applying forces and materials and effects that are useful—that do work. It troubles me when academics put such faith in their models and constructs. Literally, they invent things out of thin air. Don't*

---

[64] http://www.biocab.org/Experiment_on_Greenhouses__Effect.pdf

[65] Ibid. *Experiment on Greenhouse Effect*

[66] PhD, Theoretical Physics, Cornell University, 1992...he is currently a professor in the Department of Physics at the Rochester Institute of Technology. I'm being unfair, but ala Michael Mann's department at Pennsylvania State University (can you say Jerry Sandusky? Sandusky is a creepy pederast, but, in my opinion, Michael Mann is more despicable and a larger danger to the human race), so if Joel is in good standing there, then RIT must be a snakepit of political correctness and bullshit.

[67] http://wattsupwiththat.com/2012/07/07/new-paper-shows-negative-cloud-feed-associated-with-sam/#comment-1026875

you know how useful if would be to heat something by an average of 33°C (GHE)? Or, how useful it would be to heat something by an average of 66°C (by doubling the GHE)? Or, how useful it would be to cool something by an average of 33°C (by blocking or defusing the GHE)? However, the activists say, you can't make this powerful force of nature work on command in your living room—you must have thin layers of cold, rarefied gases to make this powerful heating engine work.

It would be easy to shut me up. Just tell me how long the GHG delay is for energy leaving our system. What is the median delay? What is the distribution of this delay? Is the delay uniform from the tropics to the poles? How does a small delay in OLWR change an average temperature in a diurnal system? How long must the delay be to contribute positive feedback: nine hours or so? Twelve? Is that how long it takes for photonic energy to leave our system? The incoming delay is something like two hours for materials (land and water) with greater thermal mass and a longer thermal time constant than air. The outgoing delay is longer than that? Prove it.
—Ken Coffman

Ken: I think history has proven that nothing can shut you up, as you have had the greenhouse effect explained to you many times by AGW skeptics and those who agree with the scientific consensus alike and it has completely failed to register. We have told you that the issue is not one of delay. It is an issue of the rate in and rate out. You have energy coming in at a certain rate and you have to have energy going out at the same rate. If you do something that decreases the rate at which energy leaves the system, like increasing greenhouse gas concentrations, then the rate at which energy leaves the system will be lower than the rate at which it enters. The system responds by increasing its temperature until the rate out again matches the rate in (which occurs because the rate at which *energy escapes is an increasing function of temperature).*

This ought not be something that a reasonably competent engineer should have a problem understanding unless they are so ruled by their ideology that they cannot understand anything that challenges it.

—Joel D. Shore

Shall we enjoy more of these types of exchanges?[68] If you want to view the world as Professor Shore sees it, go right ahead. You have my permission to be a moron. You see, I like competing with morons in the job market.

You have failed to prove the existence of back-radiation. You can't invent it, then reradiate it, then use it instead of net radiation to heat warmer air below.

—Pierre Latour, PhD, Chemistry

Back-radiation is simply the term used to describe radiation from the atmosphere that is directed back toward the Earth. I don't need to prove it, as such radiation can be (and is) measured and the fact that things with a temperature radiate is a well-established piece of physics. Why should the atmosphere be any different than everything else?

Yes, the NET effect of the radiation that the Earth's surface emits to the atmosphere and that the atmosphere emits to the surface will be cooling, not warming. However, it will be less cooling than would occur if all of the radiation from the Earth's surface escaped out into space. And, since the temperature of the Earth is determined by the balance of what it receives from the Sun and what it radiates back out into space, this reduction in the Earth's ability to cool itself results in an increase in the temperature (until radiative balance is restored).

---

[68] http://www.globalwarmingskeptics.info/attachment.php?aid=417

Ken Coffman

—Joel Shore, PhD, Theoretical Physics

At the bottom of the bottom, let's salute and appreciate fair-minded scientific explorers like Craig Bohren:

How Do Greenhouses Work?

The atmospheric science community seems to be divided into two groups. In one group are those who, when hearing the words "greenhouse effect," roll their eyes, shake, foam, and turn a delicate shade of purple while sputtering, "Greenhouses don't work that way." In the other group are those who are quite content with the term greenhouse effect. Indeed, they would probably be just as content if it were called the outhouse effect. I can afford to have a bit of sport over this because at one time or another I have belonged to both groups.

As you might expect, the term greenhouse effect, used to describe increased temperatures because of increased atmospheric carbon dioxide, arises from real or imagined connection with greenhouses. To some folks the interiors of greenhouses are warm because of (metaphorical) radiation trapping: the glass is transparent to solar radiation but opaque to infrared radiation. The glass absorbs infrared radiation from the underlying soil, is warmer than it would otherwise be, hence keeps the soil warmer than it would otherwise be. If this explanation is correct, it is reasonable to describe the effect of increasing carbon dioxide as a greenhouse effect.

Other folks, however, assert that radiation trapping has little to do with the interior warmth of a greenhouse. They are merely shelters from the wind. That is, they suppress convective heat transfer (for a discussion of the various modes of heat transfer see Chapter 8) rather than radiative heat transfer. If this is correct, the atmospheric greenhouse effect is a misnomer.

Both explanations have their adherents. Both are reasonable. What is worse, calculations can support either of the two views. After years of meditation, I have come to the conclusion that when reasonable people hold diametrically opposite views on a scientific subject, it is often that both are right—and wrong. One's conclusions depend very much on one's assumptions, both stated and unstated.

Over ten years ago (1974) a controversy about the mechanism for greenhouse warming erupted in the pages of the *Journal of Applied Meteorology*. A parallel, but nearly independent, controversy was initiated a year later by Ronald Schwiesow (now at NCAR) in the pages of *Optical Spectra*. This led to a spate of letters in this journal and in *Science*, *New Scientist*, and *Popular Science*. I am grateful to Schwiesow for passing on to me his collection of greenhouse controversy papers to add to mine.

John Kessler, a physics professor at the University of Arizona, wrote an incisive analysis of the various claims and submitted it for publication. Kessler's manuscript was rejected (for reasons I can't remember), but I have kept it all these years, and upon re-reading it, I find that is makes as much good sense now as it did then.

According to Kessler, both sides to the controversy have merit, which ought to please everyone. The relative contributions of convective suppression and radiation trapping depend very much on the greenhouse and its environment. The thicker the glass and the stiller the air, the more important is radiation trapping. The greater the wind speed and the thinner the glass, however, the less important it is. This should perhaps come as no surprise. On a cold day in a howling wind you stay inside your home, not because it traps radiation but because it shelters you from the wind. Indeed, it is almost a truism of heat transfer that the relative contributions of the various modes depend very much on circumstances. In Chapter 8 I suggest some ways to *demonstrate this.*

One of the most interesting examples of the role of convection in greenhouse temperatures is in a paper by Kirby Hanson in the Journal of Applied Meteorology (1963, Vol.2. p. 793). He states that "during January 1961 the minimum temperature averaged 2.4 F lower inside than outside" a small polyethylene-covered greenhouse. Here is an example off a greenhouse in reverse; the greenhouse suppresses mixing of warmer surrounding air with that inside it.

The role of the thickness of the glass is perhaps not so obvious. The thicker the glass, the greater the temperature difference between its outer and inner surfaces. Heat transfer from the glass to its surroundings depends on the temperature difference between the outer surface and the surrounding air. The colder this surface is, the lower the rate of heat transfer, hence the hotter it is inside the greenhouse. If you find yourself under fire for using the term greenhouse effect as shorthand for what happens in the atmosphere you need merely retort that your thick-walled greenhouse is set in a very calm spot. On the other hand, if you either snicker or fume when you hear or see the term greenhouse effect, you can draw sustenance from the fact that greenhouses are often set in windy environments and have thin walls.

THE OTHER GREENHOUSE CONTROVERSY

There is only one sure way of determining whether increased atmospheric carbon dioxide will cause average global temperatures to rise. Construct two planet Earths, identical in all respects except that on one carbon dioxide remains constant and on the other it increases with time.

Then observe the temperature changes on both planets. This project, desirable though it might be, is something that even a NASA project manager would shrink from undertaking. So we have to rely on an inferior substitute: computer modeling.

When movable type was invented the printed word became sacred, it was difficult to accept that words so beautifully displayed could be devoid of sense. Now that we have lived with books for centuries they have lost some of their sacredness, and the printed word is viewed with more skepticism. But computers are a recent innovation, and the output that spews from them tends to be regarded with the same awe once given to the printed word. Yet no matter how large the computer, no matter how many thousands of acres of virgin forest are devastated to provide paper for its maw, the results are no better than the ideas and data that went into their production.

The current conventional wisdom is that doubling carbon dioxide will cause global temperatures to rise 2 or 3 degrees C. A vocal minority, however, asserts that not only will the temperature not rise, it will decrease. All such predictions necessarily require that simplifying assumptions be made. The boundary between simplification and oversimplification is obscure, but woe to him who crosses it. That carbon dioxide concentrations in the atmosphere have been increasing seems incontrovertible. Only the consequences of this are in dispute. Why this is so is because of feedbacks that are either unknown or are swept under the rug. The best example of feedback that I have ever seen was a black box on the side of which there was a toggle switch and a sign: DO NOT TURN ON. Perhaps you could resist the temptation, but I couldn't. The result was, after some grinding and whirring from within the box, that the lid opened, a hand emerged, turned off the switch, and then withdrew back into the box. Turning on the switch resulted in a feedback that eventually turned it off. And so it is also with the atmosphere. Increasing temperatures as a result of increasing concentrations of carbon dioxide may cause something unforeseen to happen that will result in lower temperatures. One possibility, for example, might be that warming the oceans will result in more water vapor in the

*atmosphere and, as a consequence, more clouds. But this might
lead to more solar radiation scattered to space, hence lower
temperatures.*

*Because of feedbacks, some known, some unknown, some
accounted for correctly, others incorrectly, the carbon dioxide
controversy will continue to rage until the evidence is before
our very eyes. I say our eyes, but I mean someone else's eyes. It
is likely that neither you nor I will be around to see.*[69]

Well said, Craig. I don't know whether the net effect of increasing
concentrations of CO_2 in our atmosphere contribute to a net effect
of heating or cooling—and neither do you, dammit. Whatever the
net effect, it is small. Immeasurable. Irrelevant. Inconsequential.
We can agree that the extremists can't be certain of anything, so
why should we rework our civilization to suit them?

The Nature of Radiation

Radiation is a tough thing to grasp—we will be arguing about its
fundamental nature and discovering new aspects of it for the
foreseeable future. However, we can *read* the mind of Mother
Nature and understand *why* it gets employed. It's easy enough:
passive radiation is an agent of entropy.

Mother Nature does not like extremes; she is constantly trying
to average things out and has several tools in her toolbox to work
with. Is there a pressure difference between point A and point B?
Mother Nature does not like that and the larger the difference, the
harder she works to integrate the difference. Is there a density
difference between fluids at point A and point B? Mother Nature
does not like that. Is there a salinity difference between ocean water
at point A and point B? Then Mother Nature applies her physical
tools to equalize. Is there a temperature difference between point A
and point B? Radiation is one of the tools used to average out the

[69] Craig F. Bohren, *Clouds in a Glass of Beer: Simple Experiments in Atmospheric
Physics,* Dover Publications (July 10, 2001)

difference.

Most engineers don't think about radiation much—for one simple reason: it is an ineffective way of moving heat from place to place.[70] Where conduction is possible, conduction will dominate over convection. Where convection is free and easy, it will dominate over radiation. Only in situations where the radiating body is very intense and when conduction and convection are restricted (for example, in a vacuum) will radiation effects even be noticeable. If radiation was useful to us, then we'd know more about it.

Radiation can be experienced; a common example is the heat from a hot campfire that warms your face after traveling through cool air (which is not heated much). Hell, where would we be without that 5,500°C ball of flaming gas which contributes 1368 W/m^2 of Sun power to warming our planet? For passive radiation to be significant, it has to come from something hot and dense.

Does that describe the cold, rarefied atmosphere of our planet?

Coffman's Law: All coherence is transitory

This silly law (really, it's a restatement of the Second Law of Thermodynamics[71] [72]) does not always mean Mother Nature wins

---

[70] When particles are present in high temperature industrial systems (such as fluidized beds or coal furnaces) there is a distinct possibility that radiative transfer by the particles may be important.
—M. Quinn Brewster, *Thermal Radiative Transfer and Properties*, p 230
I embrace the corollary: unless emitted by a dense mass at high temperature, radiation is very difficult to measure. How can we use something this paltry and weak to get useful work done? In addition, I've already said this: I only believe in things we can measure.
[71] The entropy of an isolated system during a process always increases or, in the limiting case of a reversible process, remains constant.
[72] M. Perrin and Langevin have made a successful attempt in this direction. M. Perrin enunciates the following principle: "An isolated

right away, but she always keeps trying. She would like the atmospheric pressure to be equal on Earth and in space, but gravity fights against dissipation and rarefaction. Eventually, Mother Nature, with a patient persistence, will win this battle, but that is a long way into the future (I hope).

Here is an exchange I had with Grant Petty.

KLC: Hello Dr. Petty. I hope all is well. I'm reading through your book (Atmospheric Thermodynamics[73]) and I haven't got very far yet, but so far I haven't found anything that irritates me too much.

I have a question for you, sir.

Here's your statement on page 14:

> *In fact, left to its own devices, heat conduction will always act to reduce temperature differences.*

I wonder if you'd accept an extension of this statement as follows:

> *In fact, left to its own devices, heat convection will always act to reduce temperature differences.*

And, let's go wild and consider the third natural extension:

> *In fact, left to its own devices, thermal radiation will always act to reduce temperature differences.*

---

system never passes twice through the same state." In this form, the principle affirms that there exists a necessary order in the succession of two phenomena; that evolution takes place in a determined direction. If you prefer it, it may be thus stated: "Of two converse transformations unaccompanied by any external effect, one only is possible." For instance, two gases may diffuse themselves one in the other in constant volume, but they could not conversely separate themselves spontaneously.

—Lucien Poincaré, *The New Physics and its Evolution*

[73] *A First Course in Atmospheric Thermodynammics*, Petty, Grant W, Sundog Publishing, www.sundogpublishing.com

I can't think of an example (without external work applied) where conduction would act in a manner contrary to convection or radiation, but maybe you can think of a case. I'm truly curious about your thoughts on this.

GP: Sure. The Sun is 6000K. Radiative exchanges between the Earth and the Sun "try" to reduce the difference between those two bodies, making the Earth warmer.

Outer space is 2.7 K. Radiative exchanges between the Earth and space "try" to reduce the difference, making the Earth colder.

The actual temperature of the Earth is determined by a competition between those two exchanges, one trying to make it warmer, the other trying to make it colder. The DETAILS of the competition, and thus the ACTUAL equilibrium temperature of any component of the Earth or atmosphere, depends on the details of what's doing the absorbing and emitting at which wavelengths.

—Grant

It takes rabid application of postmodern science to dispute something so fundamental.

> *Thermal radiation is electromagnetic radiation emitted by particles of matter (molecules, atoms, ions, and electrons) as they undergo internal energy state transitions. The radiative energy produced by these transitions is usually in the ultraviolet, visible, and infrared portions of the electromagnetic spectrum. Like all forms of electromagnetic radiation, thermal radiation travels at the speed of light. Thermal radiation is also a form of heat. Heat, defined as thermal energy transfer from one body of matter to another due to a temperature difference, appears in two fundamental forms: conduction and radiation. The fundamental mechanism of energy transport in conduction is direct exchange of kinetic energy between particles of matter. In radiation, the fundamental mechanism of energy transport is by*

*electromagnetic waves (or photons) that are emitted and
absorbed by the particles of matter as they undergo energy
state transitions.*

*Like conduction, thermal energy is in harmony with the
second law of thermodynamics such that, in the absence of
work, thermal energy is radiated spontaneously from higher
temperature to lower temperature matter. Unlike conduction,
however, which requires a material path, radiative transfer
can occur between two spatially remote bodies of matter at
different temperatures even when the intervening space is a
vacuum.*[74]

However, as silly as it is, climate academics reject these simple
notions. Let's look at some of the goofball things that are said.

*It was a conceptual leap to recognise that the air itself could
also trap thermal radiation.*[75]

*Greenhouse gases in the atmosphere, especially water vapor,
"trap" (absorb and emit) some of this infrared radiation, and
keep the Earth habitably warm.*[76]

*The global system that regulates the Earth's temperature is
very complex, but many scientists believe that the increase in
temperature is caused by an increase of certain gases in the
atmosphere that trap energy that would otherwise escape into*

[74] M. Quinn Brewster, *Thermal Radiative Transfer and Properties*, Wiley-Interscience, February 1992. I owe a debt of gratitude to my colleague Alan Siddons for bringing this book and quote to my attention.
[75] IPCC Fourth Assessment Report, *Climate Change 2007: Working Group I: The Physical Science Basis, Examples of Progress in Understanding Climate Processes, The Earth's Greenhouse Effect.*
http://www.ipcc.ch/publications_and_data/ar4/wg1/en/ch1s1-4.html
[76] http://www.weatherquestions.com/What_is_infrared_radiation.htm

space.[77]

One possible answer was a change in the composition of the Earth's atmosphere. Beginning with work by Joseph Fourier in the 1820s, scientists had understood that gases in the atmosphere might trap the heat received from the Sun. As Fourier put it, energy in the form of visible light from the Sun easily penetrates the atmosphere to reach the surface and heat it up, but heat cannot so easily escape back into space. For the air absorbs invisible heat rays ("infrared radiation") rising from the surface. The warmed air radiates some of the energy back down to the surface, helping it stay warm. This was the effect that would later be called, by an inaccurate analogy, the "greenhouse effect." The equations and data available to 19th-century scientists were far too poor to allow an accurate calculation.[78]

It can take the Earth thousands of years to warm up or cool down just 1 degree when it happens naturally. In addition to recurring ice-age cycles, the Earth's climate can change due to volcanic activity, differences in plant life, changes in the amount of radiation from the Sun, and natural changes in the chemistry of the atmosphere.[79]

Most scientists agree that global warming is caused by the greenhouse effect.
Greenhouse gases in our atmosphere (CO_2, water vapour, methane etc.) absorb thermal (infrared) radiation from the

---

[77] http://www.mpcfaculty.net/mark_bishop/greenhouse.htm

[78] http://www.aip.org/history/climate/co2.htm

[79] http://science.howstuffworks.com/environmental/green-science/global-warming1.htm This quote is particularly dumb. Do you think a volcano takes thousands of years to cool the Earth after an eruption? How fast do you think the Earth cools when the sun cycles into a mode where there's less emission?

Sun. Because there are more and more greenhouse gases in our atmosphere, more heat is being absorbed. As the air heats up, the Earth's surface temperature is rising too.[80]

GHGs absorb infrared (heat) radiation. Adding GHGs to the air makes the atmosphere more opaque at infrared wavelengths where the Earth emits heat radiation. Increased infrared opacity causes emission to space to come from a higher, cooler level in the atmosphere. Thus if GHGs are increased the planet temporarily absorbs more energy from the Sun than it radiates as heat.

This energy imbalance causes the planet to warm up until energy balance is restored.[81]

Over and over, you'll read how our atmosphere can trap, block or store outgoing radiation or come across things like the following:

The mean temperature of the atmosphere is determined by the total incoming and outgoing radiation.[82]

We can certainly measure incoming and outgoing radiation and it certainly must balance, but how do we extract useful work from a radiation balance? Imagine a heat engine running off radiation aimed hither and thither. Unless there is a large mass at a high temperature, radiation is weak tea. It's more of an effect than a cause of anything. It's just plain silly.

I have to give Scorer credit because later he says this:

---

[80]

http://wiki.answers.com/Q/How_does_the_global_warming_system_work

[81] James Hansen, Congressional Testimony:
http://www.columbia.edu/~jeh1/mailings/2011/20110126_RatcliffeT estimony.pdf

[82] *Dynamics of Meteorology and Climate*, Richard S. Scorer, John Wiley and Sons, 1997, p 532

In spite of great variations in the power of the Sun there has been a large part of the Earth's surface at a temperature near enough to 10°C for it to be very suitable for life for 3000 million years. How the Earth has avoided much larger variations in climate than actually occurred is a much more important problem because the stabilizing mechanisms might overwhelm any effect due to the activities of mankind.[83]

I submit this as proof that Dr. Scorer is not completely insane.

The most popular graphic representation of madness is in Jeffrey Kiehl, John Fasullo and Kevin Trenberth's famous energy balance diagram. After Michael Mann's infamous hockey stick graph, I believe this is the second most iconic global warming chart. From an engineering point of view, this has to be one of the most blatantly dumb charts ever produced, but you'll find all kinds of defense for it from very educated people.[84]

Let's be clear: when I describe something as nonphysical, that means it represents nothing seen in nature. In other words, it's made-up. Fictional. Malarkey.

Take a look at Figure 6. Don't let your eyes glaze over; engage the gray stuff between your ears just for a few moments. Notice it shows $161 W/m^2$ of warming radiation coming from the Sun. Don't let the units of W/m^2 trouble you; it's simply calculated power per square meter. Notice it shows $333 W/m^2$ coming from atmospheric "back radiation". Hell, I could end this book right now. There are complicated explanations about how they came up with the numbers used in the energy balance diagram—explanations that include hundred-dollar words like equilibrium climate, reanalysis fields, meridional structure, improved aerosol climatologies and surface turbulent fluxes.

---

[83] Ibid. p 533
[84] http://scienceofdoom.com/2010/07/26/do-trenberth-and-kiehl-understand-the-first-law-of-thermodynamics/

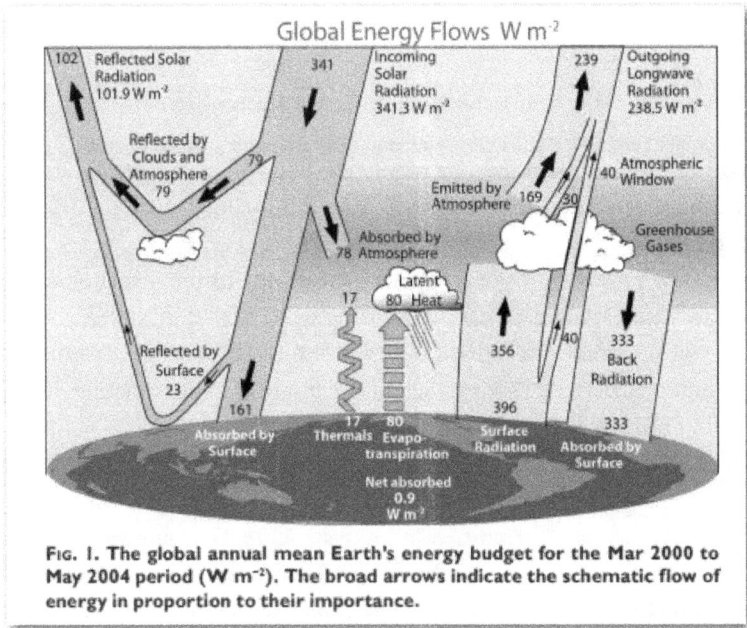

Fɪɢ. I. The global annual mean Earth's energy budget for the Mar 2000 to May 2004 period (W m⁻²). The broad arrows indicate the schematic flow of energy in proportion to their importance.

Figure 6

The Trenberth-Fasullo-Kiehl Energy Balance Schematic[85]

From this we get 33°C of additional surface warming. There is an amazing amount of work being done—the cold, thin atmosphere increases the Earth's surface temperature by around 10%.[86] Wow!

I'm not so concerned about the radiation going to and fro. The question is: how much work can the radiation do? Look at the

---

85

http://www.cgd.ucar.edu/cas/Trenberth/trenberth.papers/BAMSmar Trenberth.pdf

[86] In degrees K, the Earth's surface temperature is assumed to be 255K (-18°C) without the effect of GHGs and is measured at 287K (14°C) with GHGs. The difference is around 10%.

following two equivalent expressions:

$$a = b;$$
$$a + 1{,}000{,}000 = b + 1{,}000{,}000$$

You can make up an infinite number of equivalent equations and all will be true and accurate. What I care about is the work that can be done.

The reason they get away with this is due to the obscure nature of radiation. It's not an easy thing to grasp.

To conclude this chapter, let quote from an email from Carl Brehmer[87] to the Principia Scientific International[88] mailing list. I could not state his points better—I tip my hat to Carl.

> Two of the most frequently named late 19th century and "early 20th century" scientists who are believed to have proven the existence of a "greenhouse gas" mediated "atmospheric greenhouse effect" are John Tyndall and Svante Arrhenius; so I read through some of their papers and lectures and discovered that both inadvertently disproved the hypothesis. Let's first look at a lecture that John Tyndall delivered at the University of Cambridge, England in 1865. Prior to this lecture he had made some laboratory observations while directing "calorific rays" through 100% concentrations of water vapor and carbon dioxide. ("Calorific rays" are, of course, today called IR radiation.) This is what he said about the affect of IR radiation on "air":
>
> "Through air…the waves of ether pass without absorption, and these gases are not sensibly changed in temperature by the most powerful calorific rays."[89] The "air"

---

[87]

http://myweb.cableone.net/carlallen/Greenhouse_Effect_Research/Welcome.html

[88] http://principia-scientific.org/

[89] Tyndall, John, *On Radiation: The "Rede" lecture*, delivered in the Senate-

that he was testing contained atmospheric concentrations of the same "greenhouse gases" that are today said to be warming the atmosphere by trapping infrared radiation. This observation has, in fact, been confirmed by millions of hours of the commercial application of infrared heating, whose manufacturers assert:

*1) "Infrared energy travels at the speed of light **without heating the air it passes through**, the amount of infrared radiation absorbed by carbon dioxide, water vapor and other particles in the air typically is negligible."[90]*

*2) "Infrared heating technology by definition **does not heat up the air**, instead it targets the objects leaving the Oxygen and humidity intact."[91]*

*3) "These infrared rays pass through the air in the room and start heating any object they hit. These rays, however, **do not heat the air** of the room or area, making it more comfortable for you."[92]*

So, both John Tyndall's own laboratory testing and millions of hours of the commercial application of infrared heating have demonstrated that at atmospheric concentrations carbon dioxide does not absorb enough IR radiation to heat the air.

John Tyndall further confirmed this when said this:

*"Whenever the air is dry we are liable to daily extremes of temperature. By day in such places, the Sun's heat reaches the Earth unimpeded and renders the maximum high; by **night** on the other hand **the Earth's heat escapes***

house before the University of Cambridge, England, on Tuesday, May 16, 1865

[90] http://www.infraredheaters.com/basic.html

[91] http://www.walmart.com/ip/Amish-Portable-Infrared-Heater/12343373

[92] http://www.positivearticles.com/Article/Why-Infrared-Heaters-are-Better-than-Other-Heaters/52630

unhindered into space and renders the minimum low. Hence the difference between the maximum and minimum is greatest where the air is driest.

"In the plains of India, on the heights of the Himalaya, in Central Asia, in Australia—wherever drought reigns, we have the heat of day forcibly contrasted with the chill of night. In the Sahara itself, when the Sun's rays cease to impinge on the burning soil the temperature runs rapidly down to freezing, **because there is no vapour overhead to check the calorific drain.**"[93]

The fact is, there is just as much carbon dioxide in the air in the plains of India, on the heights of the Himalaya, in Central Asia and in Australia as there is anywhere else; yet John Tyndall observed that in these locations "there is **no vapour** overhead to check the calorific drain."

So, both in his own laboratory experiments and in his observations of the real world atmosphere John Tyndall observed that, at atmospheric concentrations, carbon dioxide does not absorb enough IR radiation to sensibly change the temperature of the air even if hit by "the most powerful calorific rays" nor does it check the atmosphere's nighttime "calorific drain."

Beyond that, John Tyndall seemed painfully unaware that even in desert climes there is more water vapor in the air than there is carbon dioxide. Take Death Valley for example. The mean yearly absolute humidity in Death Valley, in spite of its very low rain fall, is around 5 gm/m^3. This is still about 10 times the amount of carbon dioxide that is in the air, which is only about 0.5 gm/m^3. Ergo, it is not even H_2O in gaseous form, i.e., water vapor, that inhibits nighttime cooling, even in humid climates, it is H_2O in liquid form, i.e., clouds, that cause an inhibition of nighttime cooling in humid

---

[93] Ibid. Tyndall, John, *On Radiation: The "Rede" lecture.*

climates. This has been demonstrated empirically through measuring nighttime atmospheric cooling rates on clear nights vs. nighttime atmospheric cooling rates on cloudy nights. Nighttime air temperatures have been observed to drop precipitously on all clear nights regardless of the level of humidity.

This was actually an honest mistake, because as the humidity goes up so does the cloud cover generally. Therefore the belief that water vapor "traps heat" in the atmosphere started as a simple case of misattribution. The nighttime thermal action of water in liquid form, i.e., clouds, was falsely believed to be caused by the presence of water in gaseous form, i.e., water vapor. Water vapor actually creates a veritable "anti-greenhouse effect" through the well-known and thoroughly studied atmospheric phenomenon known as "moist convection." As water vapor expands, ascends and cools it releases its latent heat content; this keeps the ascending parcel of air warmer than its surroundings, which further accelerates upward convection. Convection currents, of course, cause surface level cooling, not warming. Greenhouses, in fact, work by blocking the very convection currents that water vapor helps create via this "moist convection." It is also well known that "moist convection" is the force that creates the afternoon thunderstorms, which can drop surface temperatures by 20-30°C within minutes.

So, for warm moist air to "trap heat" in the lower atmosphere it would need to somehow resist the natural forces which compel it to expand, ascend, cool and create the very convection currents that cool the surface of the Earth. Water in liquid form, i.e., clouds, on the other hand, behave differently; they do not obey the Ideal Gas Law; they do not expand, ascend and cool. Clouds, therefore, have the ability to act like a quasi-ceiling of a greenhouse at night and reduce nighttime upward convection currents; this inhibits nighttime surface cooling. This difference between the behavior of H_2O

in gaseous form vs. the behavior of H_2O in liquid form is a measurable, observable, empirical reality; it is not simply a hypothesis or a mathematical calculation.

*Now let's take a brief look at the precondition that Arrhenius placed upon his own hypothesis. He said the following in his seminal paper 'On the Influence of Carbonic Acid in the Air upon the Temperature of the Ground'.[94] "...we will suppose that the heat that is conducted to a given place on the Earth's surface or in the atmosphere in consequence of atmospheric or oceanic currents, horizontal or vertical, **remains the same in the course of the time considered**, and we will also suppose that the clouded part of the sky remains unchanged. It is only the variation of the temperature with the transparency of the air that we shall examine."*

As you can see, the precondition that he placed on his own hypothesis is that the Ideal Gas Law must cease to operate in order for his thesis to work, because when the temperature changes so do atmospheric currents! His calculations were based upon the precondition that when carbon dioxide warms up along with the rest of the air that it will not expand, ascend skyward and help create the very upward convection currents that, again, cool the surface of the Earth.

So, what we actually have from the late 19th century and early 20th century scientists is:

1) an observation that, at atmospheric levels, carbon dioxide does not absorb enough IR radiation to heat the air nor does carbon dioxide inhibit the nighttime cooling of the atmosphere. (This has been confirmed by millions of hours of the commercial application of IR heating.)

2) Plus, we have a hypothesis that asserts that if the Ideal

---

[94] Arrhenius, Svante, *On the Influence of Carbonic Acid in the Air upon the Temperature of the Ground*, Philosophical Magazine and Journal of Science Series 5, Volume 41, April 1896, pages 237-276

Gas Law didn't exist then a doubling of carbon dioxide might warm the atmosphere a few degrees.

Bravo, Carl. Well said.

ACADEMIA AND POST-MODERN SCIENCE

With respect to water vapour, you are repeating a common misconception. Water vapour is indeed the most important greenhouse gas (and no climatologist has ever disagreed). However, the amount of time that any individual water molecule is in the atmosphere (the lower part at least) is around 10 days. Thus water vapour can be considered to be in a dynamic equilibrium with the surface conditions, trace gas amounts and aerosols on time scales longer than a month. Therefore water vapour levels in the atmosphere are a feedback and not a forcing, and are always modeled as such. The reason why CO_2 (and CH_4 and CFCs and N_2O and O_3) are important is because they absorb in parts of the spectrum where water doesn't have much impact. There is a small potential for the direct forcing of water vapour by changes in irrigation patterns, but this appears to be small on the global scale.
—Gavin Schmidt[95]

THE DISCONNECT BETWEEN ivory tower academia and objective reality is very disconcerting. I introduce this chapter with a classic quotation from Gavin Schmidt—I defy you to make sense

---

[95] http://www.realclimate.org/index.php/archives/2004/12/michael-crichtons-state-of-confusion/comment-page-2/#comment-361

of it. Evaluated academically, it serves its purpose because it's complex and obtuse. In engineering, forcing and feedback are very clear. Forcings are inputs to a system and feedbacks are outputs to the system which are returned to the input and close the loop. There are hundreds of books on control theory you can read to get these concepts straight.

Sometimes one wonders if the activist is dishonest or simply stupid. For example, Al Gore throws up a chart showing CO_2 levels following temperature changes by 800 years or so and says the CO_2 *causes* the temperature change. Great. In academia, all you have to do is say things are true, couch the explanation in complex terms and wave your hands around. Viola, conventional climate wisdom is created.

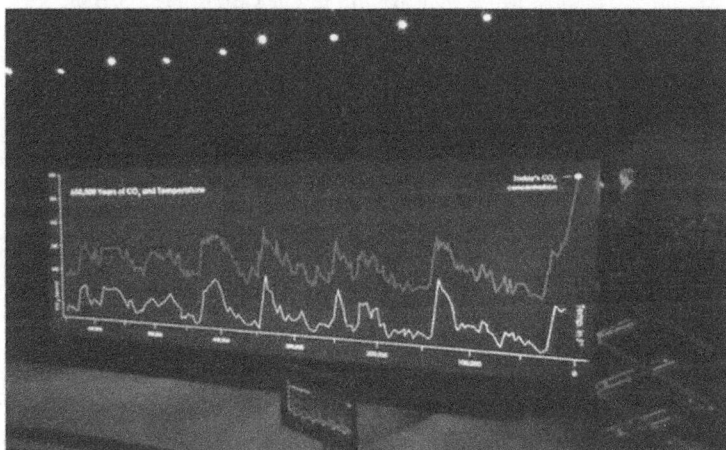

Al Gore and his Ice Core Record[96]

Figure 7

---

[96] Clip from *The Inconvenient Truth* documentary. *The relationship is actually very complicated but there is one relationship that is far more powerful than all the others and it is this. When there is more carbon dioxide, the temperature gets warmer, because it traps more heat from the sun inside.*
—Albert Gore

LITTLE CARBON DIOXIDE SUNS

I note that my box on the lapse rates was completely and utterly ignored which may explain to some extent my reaction, but I also think the science is being manipulated to put a political spin on it which for all our sakes might not be too clever in the long run.
—Peter Thorne, University of East Anglia Research Associate Professor and IPCC contributor and reviewer.

The Lack of Insight, Intuition and Common Sense Shown by Climate Activists

ONE THING YOU'LL notice when dealing with climate activists is how little the real world intrudes on their thinking. As a random example, you might ask yourself how our atmosphere, which is rarefied and mostly cooler than the ocean water, can contribute to heating ocean water? Isn't this simply common sense? If you want to heat something, don't you need something warmer than the object you're trying to heat or some other way of transferring energy? If you want to know how to heat (or cool) something, ask an engineer, not a climate activist…

It turns out that the heat transfer due to radiation, assuming that the radiating surface temperature is greater than the ambient temperature, is approximately ½ the amount of heat transferred by convection. The heat transfer contribution by Thermal Radiation is often ignored in any analysis, due to its added complexity, but may be considered as a margin of safety in the thermal design.[97]

Visualize a cubic meter of air in front of your face. It will have an average temperature. Superposition[98] is a handy principle for evaluating the effects from multiple sources or causes—you can simply take all the individual contributors and sum them up. Easy.

Now, given that CO_2 is something like 400PPM and Nitrogen, Oxygen and Argon can be rounded off to 1,000,000PPM, how can CO_2 contribute anything measurable or significant? Only if the rarefied gas is extraordinarily hot, my friends. Hence comes my genius theory of Little Carbon Dioxide Suns as described in the essay below.

Little Carbon Dioxide Suns, a Kinky Affair, Part One

If you have been following along, I figured out what the Warmists think outgoing thermal radiation does to the almighty, satanic carbon dioxide molecule. It heats it up. This probably seems like a really stupid question—I mean, what else could it do? I just

---

[97] —*Thermal Considerations for Surface Mount Layouts*, Charles Mauney http://focus.ti.com/download/trng/docs/seminar/Topic%2010%20-%20Thermal%20Design%20Consideration%20for%20Surface%20Mount%20Layouts%20.pdf

[98] In physics and systems theory, the superposition principle, also known as superposition property, states that, for all linear systems, the net response at a given place and time caused by two or more stimuli is the sum of the responses which would have been caused by each stimulus individually. So that if input A produces response X and input B produces response Y then input $(A + B)$ produces response $(X + Y)$. Source: Wikipedia

wondered if they thought it went into some mysterious higher-level hocus-pocus quantum state or something. But no, CO_2 molecules absorb IR photons and vibrate. That's fine.

The Earth radiates thermal energy. If we absorb more than we radiate, then the temperature will increase until balance is restored. We all agree on that. Please note I'm far from agreeing that the only source of heat at our surface is from solar radiation. It could be, but I don't know and neither do you.

For example, why is there a huge molten core in our Earth? Is it due to residual heat from the formation of the Earth? Is it due to nuclear decay? Is it due to eddy currents caused by solar electro-magnetic radiation? Is it a mix of these factors and others? If so, what are the relative contributions? When you nail the answers to these questions—rent a tux, you're going to Stockholm for your Nobel Prize (Physics). Don't forget to say hi to your mom while you're on camera.

Before we dig in, let's define the basic concepts for moving thermal energy around.

Conduction This is a mechanical, physical interaction between vibrating atoms. This is the easiest way to move heat from hot objects to cooler ones. Say you hold a metal spoon in a flame and burn your hand? Crap, that hurts. That's because the flame's heat was conducted to your hand. Don't try this with a plastic or wooden spoon, bozo.

Convection This is similar to conduction, except the hot atoms physically move from point A to point B and carry their heat along with them. In your furnace, some sort of heat source warms the air and a fan moves the warm air around your house. The heat stored in the air is convected: physically moved from place to place.

Radiation Heat can be transferred by electromagnetic waves (or infrared photons, depending on how you prefer to visualize the effect). You stand outside on a sunny day and the Sun warms your flesh. It's as if by magic, because you know the mostly-empty space between the Earth and Sun is cold, cold, cold, but the heat finds its way to your skin anyway. Amazing. We don't need to get too deep into this lest your eyes glaze over, but please make a note: the hotter something is, the more it radiates.

Warmists believe CO_2 affects the Earth's energy balance and creates measurable heating on the surface. That seems completely absurd to me, but that's their argument. How can this happen? They call CO_2 a heat-trapping molecule. How interesting. A heat-trapping molecule. Think about that for a minute. Wow.

If you want to talk about heat trapping, let's talk about insulation. You have heat-trapping insulation in your house, so the concept is familiar and comfortable. Have you ever seen a CO_2 rating for this insulation?

I'll wait while you run down to Home Depot and take a look.

No? I didn't think so. That's because the insulation effect is created by air trapped in the material. There's fiberglass, but that's mainly for structure—to hold the walls of the insulation apart. The structural fiberglass material doesn't conduct heat very well, so that's good. You could use steel wool in place of the fiberglass, but metal is a good conductor of heat, and you'd be a crappy insulation designer. You're useless. You're fired. Go find a government job as a climate scientist.

While you were at Home Depot, I hope you noticed something about the insulation. The thicker the insulating material, the more air it holds and the better it insulates.

You might want to think about that for a moment. Air is a gas and it's an excellent insulator. Oh boy. There is no mention of the CO_2 content of the air because it's irrelevant to the insulating property of that nasty pink stuff you're buying.

As an aside, some windows are filled with Argon—an inert gas. Some are filled with Krypton (which is a little better). What's the deal? The glass conducts heat pretty well, so a single-pane window has a low insulating value ('R' or thermal resistance value). Trap air between the panes and the insulating effect increases. But still, the air circulates between the panes and convects heat between the panes. If you used something thicker than air, this circulation would be reduced and the insulating effect would be increased.

Hence: Argon, my friends. And even better (denser) if you can afford it? Krypton.

I don't want to get too far afield. Air is an excellent insulator. Our Earth is surrounded by a relatively thin layer of it. So, heat *is* trapped. Air is an excellent heat-trapping gas and you can thank your lucky stars (or your Lucky Stars™ if you prefer) that this is so. Otherwise, we'd be broiling when exposed to the Sun and frozen solid the instant the Sun disappeared over the horizon. And, that's not all you should be thankful for.

Most of the Earth's surface is covered with water. Water conducts and convects heat pretty readily and has a massive ability to store energy—this moderates the temperature extremes we're exposed to. However, that's not all. In our environment, it thaws, freezes, evaporates and condenses. Mind-boggling amounts of energy are stored (hidden) and released in these processes. There are trip points at zero °C (freezing) and 100°C (boiling)— incredible obstacles to the temperature going above and below these points. This does not prevent ice and water vapor from existing (duh!), but it means there is a huge energy "cost" and "dividend" which helps to keep the environment friendly for us.

Let's take a look at CO_2 in the atmosphere. First of all, there is not much of it. Our atmosphere is mostly Nitrogen and Oxygen— CO_2 is about 400PPM. This means, for each 2564 molecules of air, one of them is CO_2. If CO_2 does anything measurable, it must have extraordinary power. We know of small things in nature that *do* have extraordinary power—the $e = mc^2$ relation for example, tells us the tiny atom holds massive amounts of energy. So, just because CO_2 is highly rarified does not automatically mean it can't do anything, but the burden of proof is large. Perhaps this is just me, but I'm more worried about being crushed by an elephant than by a gnat.

A warmist once mocked me—a virus is small and *it* can kill me. Ha-ha, funny. Yes, a virus is small, that's true, but one is not going to kill me, though a lot of them could. See how that works? Small things can kill you, but it takes a lot of them. Yes, I'm aware of highly potent neurotoxins. Small amounts of them can kill me. CO_2 is not a neurotoxin. Stop it, I'm bored already.

If CO_2 does something measurable to the Earth's temperature and there is very little of it, then it must have tremendous heat storage capability. That's where my concept of Little Carbon Dioxide Suns (LCDS) comes from. Now I'm very afraid. How hot do LCDS get? Hot enough to boil water ($100°C$)? I've poured boiling water on my hands and it hurts. It hurts a lot. I'm afraid.

But, wait. I've been outside when the Earth was radiating and stimulating these LCDSs. Surely they hit my skin many times. Maybe many times per second. I'm uncharred. I'm unbroiled. Some aspects of science are so goofy that only a government or academic climate scientist can understand them. This is one of those cases.

You might wonder—how can LCDS not burn me to a crisp, but be able to increase the Earth's temperature? The Earth is massive—it would take a huge amount of energy to move the temperature by any measurable amount. The answer is: the LCDS concept is nonsense. It's stupid. Insane.

The Earth radiates. How much it radiates is based on its temperature and the temperature difference between materials (like the difference in temperature between the Earth's surface and a cloud overhead).

Suppose we accept the silly idea that CO_2 is a great absorber of energy? For the effect to linger very long, the molecule must have a large thermal mass. But the thermal mass of a rarified molecule is minuscule. It is not going to hold onto its heat for very long. In other words, at night, when the outgoing IR is at a minimum, anything the CO_2 can do must be depleted. The warming effect is reset every time the atmosphere cools—the effect cannot accumulate.

I am engineer. Like all engineers, we're paid to be practical. There are many factors that contribute small effects, but my boss won't pay me to explore them. I need to handle the big things and know what's irrelevant and can be ignored. Outgoing IR radiation stimulates CO_2 molecules. However, CO_2 molecules do not float around in isolation—this energy is coupled to the rest of the atmosphere. Through conduction, the Nitrogen and Oxygen are at the same temperature and through convection this heat energy is moved around.

The Earth's temperature is integrated and is mainly based on radiation from the Sun heating the water in our oceans. There are a huge number of variables that influence the surface temperature at a specific spot—albedo, cloud cover, aerosols, ocean and wind currents. Forget all of that. The Sun heats the water. The air insulates the hot water from the cold vacuum of space. In an enclosed area (like a greenhouse or the Earth floating in space), if incoming energy is unimpeded and outgoing energy is restricted, then the temperature will increase. This is not magic. If the Sun does not come out, then the greenhouse will eventually cool down. We're not creating or destroying energy—just holding on to it for a little bit while waiting for the Sun to rise again.

That's all you need to know and despite the smokescreen of convoluted logic and mangled science of climate scientists, it's not that complex.

You don't need to invent absurd twaddle like LCDS to explain our climate.

As a final test, the warmists believe there is global warming and it is caused by man's busybody activities on the 14.5% of the Earth's surface that is useful to us. To me, that seems absurd on its face, but never mind. They believe the CO_2 component of our atmosphere is increasing monotonically (in one direction and upward) due to human industry. Okay, here's my theory and you can argue about it if you like. If the signal is global, then it will be present in any sample observed anyplace. Fine. There is a public temperature record near my hometown in southern Oregon. Grants Pass. It is easy to find online, but if you want a link so you can make your own plots, I'll provide it.

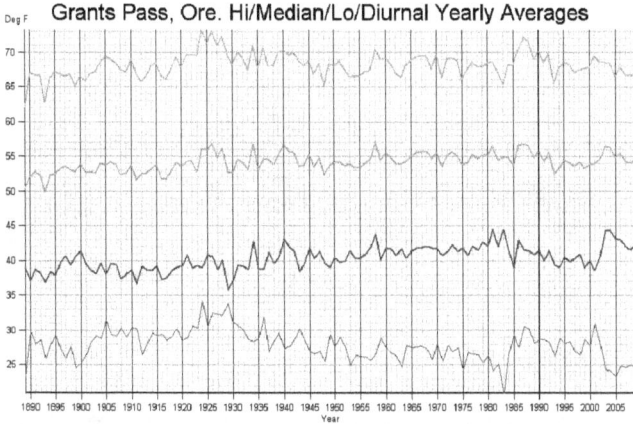

Grants Pass, Oregon, Temperature Record

Figure 8

There will probably be some Urban Heat Island (UHI) effect, but this is a small city, so let's buy the warmist argument and assume the UHI is small. This is such a simple test, you don't need to know anything extraordinary about statistics or math. Let's just look at average temperatures for January from 1928 to 2006. One record is missing (2003), so we'll just skip it. I won't bother with infilling data, we'll simply work with the data we have without "adding value". Ahem, where's the hockey stick?

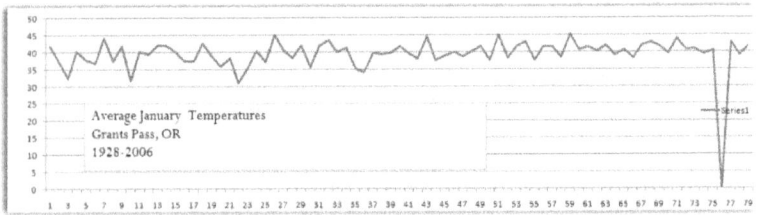

Grants Pass, Oregon, Temperature Record

Figure 9

In how many of the months does a CO_2 increase (I don't really believe this is monotonic, but we'll grant this warmist point for now) correlate with a temperature increase from year-to-year? If the warmist are right, the relation will be close to 1:1. The closer to 1:1 you get, the stronger the correlation. If the temperature variable is unrelated, then the connection will be random, at or around 50%. It would be interesting if there was anti-correlation— in other words, a strong link, but an inverse one.

What does the data tell us?

For the 77 years in question, the relation is 48%. In other words, there is no correlation.

What could I be doing wrong?

Okay, to anticipate a warmist complaint, let's just take months where the human-caused CO_2 activity kicks into overdrive—and start our test at 1940. That's 65 years of data. The relation is 49%.

I can hear the caterwauling now. *You can't extract a global climate signal from a sample size of one!* You need to be a climate scientist to make sure the samples are selected and massaged correctly to *see* the signal. Okay, I want you to listen to me carefully. If a signal is not present in an individual sample, then adding more samples does not increase accuracy! If you know of some reason the sample is corrupted or otherwise in error, that's a different matter (thank you to the ad hoc team of weather station auditors).

I will provide the spreadsheet for anyone who wants to take a look at it. It's crude and stupid. But, that's all it takes to debunk the nonsense spread by lefty-activist climate scientists from government and academia. It's amazing they had such an impact for so long when it is so easy to demolish their silliness.

But, that's the crazy world we live in, my friends.

LITTLE CARBON DIOXIDE SUNS—PART 2

When people expect problems, they're more likely to find them.
—Lars Perner, PhD, Assistant Professor of Clinical Marketing at Marshall School of Business—University of Southern California

AS A SEQUEL to my first essay about the amazing theory of Little Carbon Dioxide Suns, I wrote Part 2, which follows below…

If you wish to think critically, you must be willing to confront the fact that what you believe might be wrong. Just having a cohesive storyline means nothing—all a cohesive storyline means is you've created a scenario that *might* be correct. It is certainly incorrect to assume that because something is complicated or explained in a complicated way—that it is true. Perhaps this is the biggest difference between engineering and the ivory tower academic world. As an engineer, you are confronted with hard facts, theories you can prove for yourself, repeatable experiments and verifiable data.

If you want to store thermal energy—you need materials with large thermal masses exposed to incoming energy. In our world, in order of significance, these are things like water (the stuff that covers 71% of the Earth's surface), Earth, and the main constituents of our dense atmosphere (N_2, O_2 and Argon). You can't trap or store very much thermal energy in things that are rarified, like CO_2, empty space (the farther you get from the Earth's surface, the more our atmosphere becomes like empty space) and radiation. Radiation is. In our atmosphere, you can't trap it. If things with large thermal masses have a temperature, then they will radiate. The average radiation depends on the average temperature. The length of time of the radiation depends on the thermal capacity. By modulating the temperature of a thermal mass, you can modulate the radiation.

Before we start arguing—yet again—about the human-caused climate change hoax, let me remind you of an important fact. Humans are pattern-seeking animals. That's why, when we look up at the stars, we see bulls (Taurus), bears (Ursa Major and Minor) and sword-bearing soldiers (Orion). When we look at clouds in the sky, we see things like fluffy bunnies and our dog's smiling face. Intellectually, you know these projections have nothing to do with the reality of clouds and stars—at least, I hope you do. We project meaning onto chaos.

If you intend to think about things objectively, projection is a human trait you should always be wary of. Before I knew anything about climate science, I had a bias—I did not believe progressive activists could be trusted to draw conclusions from science without interjecting their worldview and philosophy. Staring into a tangled web of data and observations and theory, they would project their preconceived notions—their certainty that man is a plague on the planet and a danger to the beautiful balance of nature. Now that I know more about the dismal state of modern climate science, I can comfortably report that my rude presumptions were completely confirmed.

Okay, can we argue about climate change now? On occasion, for fun (you should try it sometime) I interact with the activists at RealClimate. If you've been paying any attention at all, you'll know they're not a very happy team right now due to the Climategate email and data leak, but never mind that. I understand most of the arguments the "warmists" use to justify the assertion that humans cause dangerous global warming with greenhouse gas emissions. I don't agree with them (and think many of their ideas are laughable nonsense), but I understand their arguments.

However, there are a few areas where they lose me. One is what exactly they think the excited CO_2 molecule does when massaged by outgoing IR radiation.

First of all, just to make sure we're all on the same page, the Earth is bathed in solar radiation which, when passing through clear air, arrives relatively unimpeded at the Earth's surface. The Sun warms the land and water and radiates back into space at a lower wavelength. It's an amazing characteristic of our atmosphere that our radiation is rich in wavelengths where CO_2 is resonant—thus narrow bands of energy are absorbed. So, in the minds of the Warmists, CO_2 becomes an obsession. This innocuous gas, so vital to plant life, becomes an evil, devil-gas emission spewed out by humans to the detriment of our biosphere.

We have outgoing IR radiation and some of it is absorbed by the CO_2 molecule. But then what? I'm truly curious and baffled by what the Warmists think the CO_2 does.

What *can* a CO_2 molecule do? It can get hot (which is a mechanical effect: vibration) or it can take on a higher quantum state. Or, am I missing something?

I don't know everything.

My train of thought is tickled by the wise ones at RealClimate:

In comment 155 *Completely Fed Up* says (on 8 March 2010 at 4:05 AM):

I can give you the proof [of global warming science] in three sentences, Mike:

1) CO_2 traps IR but not visible light.
2) The Sun shines bright in the visible, the Earth shines in the IR.
3) We have produced stupendous amounts of CO_2 in burning fossil fuels

Cool! Here's my question:

In comment 172 *Ken Coffman* says (on 8 March 2010 at 9:56 AM):

1) CO_2 traps IR but not visible light.

I often wonder what you guys think the almighty CO_2 molecule does with absorbed IR. Sure, it generally absorbs IR from one vector (Earth outward) and re-radiates in all three dimensions, but surely you realize most of the CO_2 molecule's temperature is due to mechanical connection to the rest of the atmosphere. Our average temperature determines what we radiate, but the dominant factor in our Earth's average temperature comes from the Sun heating water which in turn heats nitrogen and oxygen. Heat is most readily moved around by conduction and convection (and this includes the most prominent method of heating CO_2). You guys get this, right?

And this earned responses 176, 177, 179, 182 and 183…

Comment 176 and 177 *Completely Fed Up* says (on 8 March 2010 at 10:23 AM)

I wonder what you think your blanket does with all that absorbed heat from your body.

PS, what does this mean? "but surely you realize most of the CO_2 molecule's temperature is due to mechanical connection to the rest of the atmosphere."

There's no mechanical connection. A gas is not a linked system.

Comment 179 *Doug Bostrom* says (on 8 March 2010 at 10:31 AM):

Start here: AIP Climate History[99]

Comment 182 *Eli Rabett* says (on 8 March 2010 at 10:45 AM):

Ken, you really need to read the simplest explanation of the greenhouse effect.[100] Seriously, even your mom could understand it.

Comment 183 *Ray Ladbury* says (8 March 2010 at 10:46 AM):

---

[99] http://www.aip.org/history/climate/
[100] Ibid

Ken Coffman says "I often wonder what you guys think the almighty CO_2 molecule does with absorbed IR. Sure, it generally absorbs IR from one vector (Earth outward) and re-radiates in all three dimensions…"

RL: Strike one! Most CO_2 relaxes in the troposphere via collisions with other molecules (mostly nitrogen and oxygen).

Ken: "…but surely you realize most of the CO_2 molecule's temperature is due to mechanical connection to the rest of the atmosphere."

RL: Steerike two! When the excited CO_2 molecule couples to the rest of the atmosphere, the energy flow is predominantly to the atmosphere.

Ken: "Our average temperature determines what we radiate, but the dominant factor in our Earth's average temperature comes from the Sun heating water which in turn heats nitrogen and oxygen. Heat is most readily moved around by conduction and convection (and this includes the most prominent method of heating CO_2)."

RL: Steeeerike three! Yer out! Ken, Dude, this is wrong in oh so many ways. All of the heat transfer from Earth's climate system is due to radiation, or if you disagree, do please explain how convection or conduction would remove heat from Earth into the inky blackness of space.

Wow, what a great response! I don't know the real identity of *Completely Fed Up* and *Bob,* but the others are well-known folks in the climate fraudisphere. First of all, let's note that none of these eminent folks actually answered the question, but they offered up some links—we'll take a look later. But first, let's look at my response to Mr. Ladbury…

Comment 191 *Ken Coffman* says (on 8 March 2010 at 12:33 PM):

I really appreciate you guys helping to make my point.

Ken Coffman says "I often wonder what you guys think the

almighty CO_2 molecule does with absorbed IR. Sure, it generally absorbs IR from one vector (Earth outward) and re-radiates in all three dimensions…"

RL: Strike one! Most CO_2 relaxes in the troposphere via collisions with other molecules (mostly nitrogen and oxygen).

KLC: Right, exactly. And this is a bidirectional process—most of the heat of all CO_2 molecules is caused by colliding with nitrogen and oxygen and molecules like water vapor.

Ken: "…but surely you realize most of the CO_2 molecule's temperature is due to mechanical connection to the rest of the atmosphere."

RL: Steerike two! When the excited CO_2 molecule couples to the rest of the atmosphere, the energy flow is predominantly to the atmosphere.

KLC: Exactly so. And when the CO_2 molecule's neighbors are warmer, then they couple heat to the CO_2 and that's where CO_2 gets most of its energy.

Ken: "Our average temperature determines what we radiate, but the dominant factor in our Earth's average temperature comes from the Sun heating water which in turn heats nitrogen and oxygen. Heat is most readily moved around by conduction and convection (and this includes the most prominent method of heating CO_2)."

RL: Steeeerike three! Yer out! Ken, Dude, this is wrong in oh so many ways. All of the heat transfer from Earth's climate system is due to radiation, or if you disagree, do please explain how convection or conduction would remove heat from Earth into the inky blackness of space.

KLC: I'm well aware of our hot water bottle floating in space—all energy into and out of the system is via radiation. But the radiation is caused by our atmospheric temperature—everything all summed together (don't you love superposition? I do.) It's very interesting that IR absorbed by CO_2 is a factor, but that factor is very small.

I agree with the insulating blanket analogy mentioned above.

The cozy sleeping bag traps warm air—radiation has very little to do with my comfort. My body heats the air via conduction (and to a smaller degree by convection and to an even smaller degree by radiation) and the foam or down trap the warm air. Take away all of the CO_2 in my sleeping bag system and I won't even notice.

Now things are really getting good, an infamous key member of the "Hockey Team", Gavin Schmidt, checks in. Oh goody!

[**Response:** This is worthy of posting only to serve an example to the rest of us. Someone who thinks that the 'greenhouse gas is like a blanket' analogy is a statement that blankets keep us warm because of the radiative effect of CO_2 is beyond parody (and reason). Please do not respond to this post, and Ken, please note that all further discussion of your confusion is OT. — gavin]

And, to top it off, the notorious Internet troll Mark A. York checks in. As usual, he contributes a bit of snark to the conversation, but no real content, but, for fun, I'll include his comment.

Comment 193 *Mark A. York* says (on 8 March 2010 at 1:06 PM):

. . .

Yes he always says this. Of course, the whole forcing and feedback scenario is lost on him too. I know this from experience. Sadly. We have publishing disagreements as well. Coffman uses the same model there: vanity. Nice job, Ray Ladbury.

And, finally, in Comment 203 *Bob* says (on 8 March 2010 at 2:59 PM):

Okay, Ken Coffman looks like a poster child for the problems we're facing. First, he comes to a site which is run and frequented by professional scientists, and presumes to tell them how science (a very basic and accepted aspect of science) works.

When he is corrected, he misunderstands the corrections, won't accept them, digs in his heels, and again points out to everyone how he is right and they are wrong.

You just can't make this sort of thing up, and I'm not sure how you reach that sort of person, or if you even can.

But I'll fall back on a previous complaint. Most people's (Americans?) foundation in math and science is so abysmally weak that they have no chance. They must trust scientists, and that trust is being poisoned (although Ken seems to have some grasp of the science, but is confusing himself with the little that he understands).

Which I guess is my real point. I love RC, and it serves a good purpose, making the science and the scientists available to the public, but I don't think "explaining the science to Joe Sixpack" is ever going to work, because he not only has no chance of understanding, but he's arrogant enough to think that what he believes is true, and he trusts himself and various pundits more than scientists.

[**Response:** These are not universal characteristics of lay people—and in fact I find the exact opposite. Do not confuse the often strident and confused people who show up in blog comments with the much larger populace. —gavin]

But, on a good note—despite the doom and gloom of my earlier post about the lengthy phases we'll go through to actually achieve climate action, I do think people will return to trusting science as temperatures rise and irrefutable visible effects (as opposed to temperature measurements, which the propaganda has succeeded in miring in that same distrust) continue to manifest.

The thing is: no matter how much the anti-crowd plays this game, they are going to lose, and when the degree of their deception becomes obvious, the backlash will probably be fierce. People are not going to like having been duped on something so important. They will blame big oil, and the media, and maybe even scientists for not forcing them to see through the lies.

The only people they won't blame are themselves.

Wow, I'm a poster child of scientific ignorance—strident and confused. Groovy. Okay, note none of these eminences actually directly answered my question. But, they did offer up some links, so let's see if the answer can be found there.

First we'll look at Eli Rabett's link: The simplest explanation of the greenhouse effect.[101]

This is a simple explanation of how warmists think the climate system works, but it doesn't explain what they think happens to CO_2 at the molecular level.

And, some snark? Even my mother could understand Rabett's explanation? Here's what mom [RIP] had to say...

> *All I know is that there have been extreme temperature fluctuations for billions of years going from most of the north covered with ice, to jungles. And after the winter we had here, let's bring on a little warming.*
> —Mama

With all due respect, Eli, I find mom's opinion to be wiser than yours.

But, there was another link offered. AIP Climate History[102]

Ah, this is a treatise by Dr. Spencer Weart. I've read his history of climate science before and really enjoyed it. He provides an outstanding, interesting and very-readable history of the human-caused global warming movement. So, I'll take a run back through it and see if there is an answer to my question. Hold on, I'll be back soon.

The main body of his book does not address my question, but digging deeper—in his essay Simple Models of Climate,[103] he does.

But as Tyndall found, even a trace of CO_2, no more than it took to fill a bottle in his laboratory, is almost opaque to heat radiation. Thus a good part of the radiation that rises from the surface is absorbed by CO_2 in the middle levels of the atmosphere. Its energy transfers into the air itself rather than escaping directly into space. Not only is the air thus warmed, but also some of the

---

[101] http://www.aip.org/history/climate/
[102] Ibid
[103] http://www.aip.org/history/climate/simple.htm

energy trapped there is radiated back to the surface, warming it further.

—Spencer Weart, *Simple Models of Climate*

Okay, there's nothing mysterious here, the CO_2 heats up. Very good.

As I said, I admire Dr. Weart's work. One thing I like about it: he doesn't play the activist games, he simply lays out the basis of current climate change theory without blatantly agreeing or disagreeing with it. This made me wonder where his personal opinion falls.

KLC: I have a sense you put yourself above the fray somewhat and simply report on the milestones and developments of climate science. To be clear, from your research and analysis, do you agree with the "warmist" position that humans are influencing climate in a measurable (and hazardous) way? Or, are you more agnostic in your conclusions?

Dr. Weart: As a historian, I am only reporting what has been happening in the scientific community, I hope in a neutral and objective fashion: this is what the journal articles say. As someone who long ago got a PhD in physics and who has been following closely the climate science publications since the late 1980s, of course I have formed my own opinions. It was back in the late 1990s that I became convinced that there is a high probability—not a certainty, but a high probability—that the increase of greenhouse gases in the atmosphere (which nobody denies is happening and is caused by humans) will raise global temperature by three degrees, give or take a degree, by late in this century, and that this will have many very harmful impacts. There is maybe a 5-10% chance that this won't happen. There is maybe a 5-10% chance that the warming will be much, much worse.

So, he buys the activist argument. I disagree, but that's cool.

Now I understand the warmist argument, but I think the idea that the CO_2 heats up is a problem for the warmists. As an aside, the perceptive will note what I consider a contradiction in the comments above.

Most CO_2 relaxes in the troposphere via collisions with other molecules (mostly nitrogen and oxygen).

—Ray Ladbury

There's no mechanical connection. A gas is not a linked system.

—All Fed Up

Perhaps I'm confused about terminology, but CO_2 molecules do not float around in isolation. George Smith puts this better than me:

Well, guess what happens in the atmosphere when GHG [Green House Gas] molecules, e.g. CO_2, absorb specific photon energies in the 13.5 to 16.5 micron range. That energy becomes thermalized, and transmitted to the ordinary atmospheric gas molecules; which eventually radiate a thermal continuum LWIR spectrum based on the temperature of the atmosphere; not the temperature of the original emitting surface.

—George E. Smith in a post at WattsUpWithThat

The thermal state of a CO_2 molecule certainly is "linked" to the state of nearby molecules—mostly Nitrogen (N_2) and Oxygen (O_2). And, this process is completely bidirectional. CO_2 can heat or cool the atmosphere and the atmosphere can cool the CO_2.

Now, back to Dr. Weart...

Dr. Weart: Nobody who understands the basic physics will disagree with the following: when an infrared photon hits a CO_2 molecule, there is a certain chance, depending on the wavelength, that the molecule will absorb the photon's energy and jump to a higher quantum level (these are mainly quanta of vibrational and/or rotational energy, by the way). The energy may then be re-emitted as one or more photons in a random direction. Or before this happens, the CO_2 molecule may bump against another air molecule and the extra energy will be transferred into kinetic energy: both molecules will move faster, which is to say, that part of the atmosphere gets hotter.

And my response...

KLC: We all agree: unleashed CO_2 changes our radiative balance, but we argue over the magnitude of the effect.

I imagine a CO_2 molecule moving around in our atmosphere. It has a temperature. It is not uncoupled. It might be warm from absorbing a photon. It might cool by emitting a photon. Its heat might couple to another molecule it interacts with, but this is a bidirectional process. I think the Sun warms the water, the water warms the air, and the relatively huge volume of air cools a hot CO_2 molecule and warms a cool CO_2 molecule. We radiate, but the fully integrated radiation is determined by our fully integrated temperature and there are factors, both known and unknown to us now, that are much more significant than rarefied CO_2.

Mission accomplished—I understand what the warmist community thinks happens to a CO_2 molecule.

I find this topic interesting and worthy of study. Regardless of what any one "expert" might tell you, if you have a three-digit IQ, you can understand the arguments and judge for yourself their merit. If you want to believe we humans, running around and industriously doing our thing on 14.5% of the Earth's surface, can move the climate one way or another, then you have my blessing.

For me?

I think I'll stick with my mom on this topic.

CLIMATE WISDOM FROM BEEVILLE, TX

For about a microsecond, there was brouhaha about a little girl named Julisa Castillo from Beeville, Texas who won an award for her anti-human-caused global warming science project. This story went viral around the world on the Internet. As it turned out, the award was a fake created by her father to make her feel good. I think we can all agree: the sentiment was nice, but the plan was terminally stupid.

Here's the newspaper article as it was originally published:

R.A. Hall Elementary School fourth-grader Julisa Castillo has been named junior division champion for the 2010 National Science Fair.

Her project, "Disproving Global Warming," beat more than 50,000 other projects submitted by students from all over the U.S.

Julisa originally entered her project in her school science fair before sending it to the National Science Foundation (NSF) to be judged at the national level.

The NSF panel of judges included former U.S. Vice President Al Gore, 14 recipients of the President's National Medal of Science, and four former astronauts.

"Before she sent it off, she just had to add more details, citations for her research, and the amount of hours she spent working on it," said Julisa's father, J.R. Castillo.

In addition to a plaque, trophy and medal, Julisa has won an all-expenses-paid trip to Space Camp at the U.S. Space & Rocket Center in Huntsville, Ala., which she plans to attend this summer.[104]

Putting aside her dad's well-intentioned, but ultimately stupid scheme, let's take a look at her work, shall we? According to a MySouTex newspaper article[105], here's what she concluded from her study:

There is not enough evidence to prove global warming is occurring," fourth-grader Julisa Raquel Castillo concluded in a science project she entered in the campus' annual science fair

---

[104] http://www.mysoutex.com/pages/full_story_landing/push?article-R-A-+Hall+fourth-grader+is+science+national+champion%20&id=7801690&instance=landing_news_lead_story

[105]   http://www.mysoutex.com/view/full_story/5061446/article-Conclusion--%E2%80%98pretty-creative%E2%80%99?

on Tuesday.

Julisa studied temperatures in Beeville for the past 109 years to develop her conclusion.

She researched online data basis of the National Oceanic and Atmospheric Administration, or NOAA, the National Weather Service, and checked out books on climate change at the Joe Barnhart Bee County Library.

Her findings:

- *Temperatures rose and fell from 1900 to 1950.*
- *Temperatures in Beeville cooled down over a 20-year period beginning in 1955 and ending in 1975.*
- *Since 2001, temperatures in Beeville have grown cooler year after year.*

I agree with her point: if a signal is truly global, then it will be present in all samples. If it is not present in some samples, it's not global and the situation is ripe for the Warmistas to cherry pick samples and prove whatever they like. This universality will be true unless there is some problem with the data caused by the instrumentation or the sample location has some unusual source of noise, or there is some other problem. If there is a problem, then that data should be discarded. There is no validity to combining noisy databases. I know we're supposed to put our faith in the climate analysis experts who mix and adjust the data until it looks right to them. Nonsense, I say. Pick a record that is as clean as possible and examine all you want. If there is a signal, you'll find it. For me (and Julisa), we looked at temperature data and noticed it goes up and it goes down and there is no correlation to the CO_2 "signal."

I know the progressives like to wear their holier-than-thou self-righteous attitude like a fashion statement. They like to think of themselves as brave warriors saving the world from the evils of human activity. But, the thing is, if you really want to see courage, try making a case against conventional global-warming wisdom. Study whatever data you can find and draw a conclusion, then stand by it. *That* takes guts.

I can only imagine the peer pressure and teaching staff "guidance" that Julisa experienced. I salute her courage in standing up against the "settled science" of global warming.

In fact, I appreciate her courage so much—I made up an award certificate and sent her a $100 bill she can save for the trip she'd reportedly like to take to Space Camp. In addition, I sent her a copy of A. W. Montford's delicious book *The Hockey Stick Illusion*.

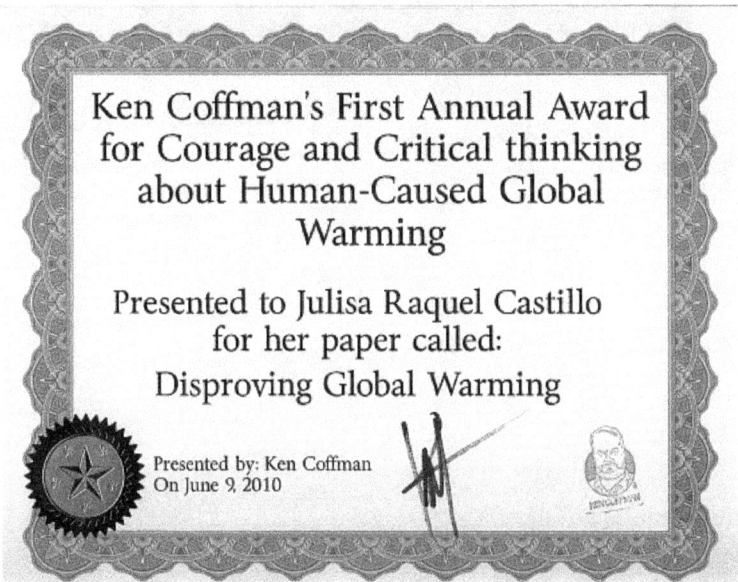

Ken Coffman's First Annual Award for Courage and Critical thinking about Human-Caused Global Warming

Presented to Julisa Raquel Castillo
for her paper called:
Disproving Global Warming

Presented by: Ken Coffman
On June 9, 2010

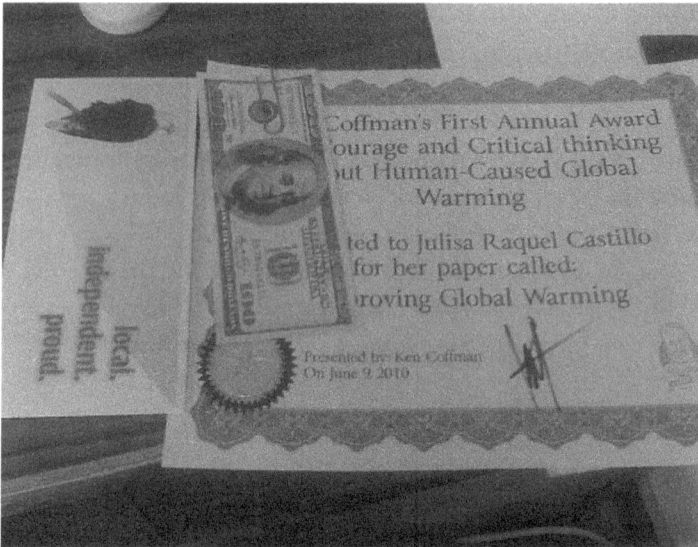

I'm pleased and honored to call Julisa my friend.

Dear Ken Coffman,
Thank you for the 100
dollars and certificate
Sorry it took this
long for a thank you.
Once again thanks.

your friend,
Julisa Castillo

A CONVERSATION WITH AN INFRARED RADIATION EXPERT

I find myself in the strange position of being very skeptical of the quality of all present reconstructions, yet sounding like a pro greenhouse zealot here!
—Keith Briffa, Deputy Director, University of East Anglia, Climatic Research Unit

An Exchange Between Ken Coffman and Mikael Cronholm—with Commentary[106]

WHILE CLICKING AROUND on the Internet, I found an outstanding paper called *Thermodynamics of Furnace Tubes—Killing Popular Myths about Furnace Tube Temperature Measurement*[107] written by Mikael Cronholm. The paper was clever and wise—and made a lot of sense. Clearly Mikael knows a lot about infrared radiation and

---

[106] http://wattsupwiththat.com/2011/02/13/a-conversation-with-an-infrared-radiation-expert/

[107]

http://flircms.com/uploadedFiles/Thermography_USA/Industries/OGI/2009-090%20Cronholm.pdf

I'm a guy with questions. A match made in heaven?

We exchanged emails. I want to be clear about this—Mikael corrected some of my wrong ideas about IR. I'll repeat that for the slow-witted. Some of my ideas about infrared radiation were wrong. I am considered a hard-headed, stubborn old guy and that's completely true. However, I want to learn and I can be taught, but not by knuckleheads spewing nonsense and not by authoritarians sitting on thrones and tossing out insults and edicts.

Ken L. Coffman: Hello Mikael. I found your paper called *Thermodynamics of Furnace Tubes* and I found it very informative, practical and interesting. I hope you'll bear with me while I ask a couple of dumb questions. I am an electrical engineer, so I have some knowledge about thermodynamics of conduction and convection, but not so much about IR radiation. In return for your time, I would be happy to make a donation to the charity of your choice.

If I take an inexpensive IR thermometer outside, point it at the sky and get a temperature reading of minus 25°C, what am I actually measuring? Is there anything valid about doing this?

Mikael Cronholm: Just as a matter of curiosity, how did you find my paper? I checked your website and I guess this has to do with the Dragon, no? If you want to make a donation I would be happy to receive that book. If you can, my postal address is at the bottom. I don't follow the debate more than casually, but I am a bit skeptical to all the research that is done on climate change—it seems that the models are continuously adjusted to fit the inputs, so that you get the wanted outputs...and they argue "so many scientists agree with this and that"...well, science is not a democracy...

About radiation, then. There is more to this than meets the eye. Literally!

Looking at the sky with an infrared radiometer you would read what is termed "apparent temperature" (if the instrument is set to emissivity 1 and the distance setting is zero, provided the instrument has any compensation). Your instrument is then receiving the same radiation as a blackbody would do if it had a temperature of -25°C, if that is what you measure. It is a quasi-temperature of sorts, because you don't really measure on a particular object in any particular place, but a combination of radiation, where that from outer space is the lowest, close to absolute zero, and the immediate atmosphere closest to you is the warmest. (I have once measured -96°C on the sky at 0°C ground temperature.) What we have to realize though, is that temperature can never be directly measured. We measure the height of a liquid in a common thermometer, a voltage in a thermocouple, etc, and then it is calibrated using the zeroth law of thermodynamics and assuming equilibrium with the device and the reference.

KLC: Global warming (greenhouse gas) theory depends on atmospheric CO_2 molecules absorbing IR radiation and "back radiating" this energy back toward the Earth. If you look at the notorious Ternberth/Keihl energy balance schematic,[108] you see the back radiation is determined to be very significant—more than $300W/m^2$. From your point of view as an IR expert, does this aspect of the global warming theory make any sense?

MC: The paper you sent me mentions Stefan-Boltzmann's law, but it does not talk about Planck's law, which is necessary to understand what is happening spectrally. I suggest you read up on Planck and Stefan-Boltzmann at Wikipedia or something. Wien's law would be beneficial as well—they are all connected.

Planck's law describes the distribution of radiated power from a blackbody over wavelength. You end up with a curve for each blackbody temperature. The Sun is almost a blackbody, so it follows Planck quite well, and it has a peak at about 480nm, right in the

---

[108] As shown in Figure 1 of this paper:
http://www.cgd.ucar.edu/cas/Trenberth/trenberth.papers/TFK_bams09.pdf)

middle of visual (Wien's law determines that). The solar spectrum is slightly modified as it passes through the atmosphere, but still pretty close to Planckian. When the radiation hits the ground, the absorbed part heats it. The re-radiated power is going to have a different spectral distribution, with a peak around 10μm (micrometer). Assuming blackbody radiation it would also follow Planck's law.

S-B's law is in principle the integral of Planck from zero to infinity wavelength. Instruments do not have equal response from zero to infinity, but they are calibrated against blackbodies, and whatever signal they output is considered to mean the temperature of the blackbody. And so on for a number of blackbodies until you have a calibration curve that can be fitted for conversion in the instrument.

That means that the instrument can only measure correctly on targets that are either blackbodies, or greybodies with a spectral distribution looking like a Planck curve, but at a known offset. That offset is emissivity, the epsilon in your S-B equation in that paper. It is defined as the ratio of the radiation from the greybody to that of the blackbody, both at the same temperature (and wave length, and angle...). Some targets will not be Planckian, but have a spectral distribution that is different. If you want to measure temperature of those, you need to measure the emissivity with the same instrument and at a temperature reasonably close to the one you will measure on the target later.

So, of course, the whole principle behind the greenhouse effect is that shorter wavelengths from the Sun penetrates the atmosphere easily, whereas the re-radiated power—being at a longer wavelength—is reflected back at a higher degree. I have no dispute about that fact. It is reasonable. So I think the Figure 1 you refer to is correct in principle. My immediate question is raised regarding the numbers in there though. The remaining 0.9 W/m^2 seems awfully close to what I would assume to be the inaccuracies in the numbers input to calculate it. You are balancing on a very thin knife's edge with such big numbers as inputs for reaching such a small one. An error of ±0.5% on each measurement would potentially throw it off quite a bit, in the worst case. But I don't know what they use to measure this, only that all the instruments I use have much less accuracy than that. But with long integration times—well, maybe—but there may be an issue there.

KLC: I am interested in some rather expensive thermopile-based radiation detectors called pyrgeometers[109]

If a piece of equipment like this is pointed into the nighttime sky and reads something like $300W/m^2$ of downwelling IR radiation, what is it actually measuring? If I built a test rig from IR-emitting lightbulbs calibrated to emit $300W/m^2$ and placed this over the pyrgeometers, would I get the same reading?

MC: "What is it actually measuring?" Well, probably a voltage from those thermopiles—and that signal has to be calibrated to a bunch of blackbody reference sources to covert it either to temperature or blackbody equivalent radiation.

Your experiment will fail, though! If you want to do something like that, you have to look at a target emitting a blackbody equivalent spectrum, which is what the instrument should be calibrated to. IR light bulbs emitting $300W/m^2$ is simply impossible, because $300W/m^2$ corresponds to a very low temperature! Use S-B's law and try it yourself. Like this: room temp, $20°C = 293K$. The radiated power from that is 293K raised to the power of 4. Then multiply with sigma, the constant in S-B's law, which is $5.67*10^{-8}$, and you get $419 W/m^2$ or something like that, it varies with how many decimals you use for absolute zero when you convert to Kelvin. For $300 W/m^2$ radiation I get $-23.4°C$ at $300 W/m^2$ when I calculate it (yes, minus!). Pretty cool light bulb.

I don't know what your point is with that experiment, but if it

---

[109] An example is the KippZonen CGR 3 instrument:
http://www.kippzonen.com/?product/16132/CGR+3.aspx).

is to check their calibration you need a lot more sophisticated blackbody reference sources if you want to do it at that temperature. But you could do a test at room temperature though. Just build a spherical object with the inside painted with flat black paint, make a small hole in it, just big enough for your sensor, and measure the temperature inside that sphere with a thermocouple, on the surface. Keep it in a stable room temperature at a steady state as well as you can and convert the temperature to radiation using S-B's law. You should get the same as the instrument. Any difference will be attributable to inaccuracy in the thermocouple you use and/or the tested instrument. Remember that raising to the power of 4 exaggerates errors in the input a lot!

I hope I have been able to clarify things a little bit, or at least caused some creative confusion. When I teach thermography I find that the more you learn the more confused you get, but on a higher level. Every question answered raises a few more, which grows the confusion exponentially. It makes the subject interesting, though. Let me know if you need any more help with your project!

KLC: I found your paper because one of the FLIR divisions is local and I was searching their site for reference information about IR radiation. I know what a 100W IR lamp feels like because I have one in my bathroom. If someone tells me there is $300W/m^2$ of IR power coming from space, and I hold out my hand—I expect to feel it. What am I missing?

MC: Yeah, you put your hand in front of a 100W bulb, but how big is your hand—not a square meter, I'm sure. It is per area unit, that is one thing you are missing. The 100W of the bulb is the electrical power consumption, not the emitted power of the visual light from it. That's why florescent energy-saving lamps as opposed to incandescent bulbs give much more visual light per electrical Watt, because they limit the radiation to the visual part of the spectrum and lose less in the IR, which we cannot see anyway. The body absorbs both IR and visual, but a little less visual.

And, here is the other clue. Your light bulb radiation in your bathroom is added to that of the room itself, which is 419 W/m^2, if the room is 20°C. Your 300 W/m^2 from space is only that. You will feel those 300 W/m^2, sure. It will feel like -25°C radiating towards your hand. But you don't feel that cold because your hand is in warmer air, receiving heat (or losing less) from there too.

Actually, we cannot really feel temperature—that is a misconception. Our bodies feel heat flow rate and adjust the temperature accordingly. It is only the hypothalamus inside the brain that really has constant temperature. If you are standing nude in your bathroom, your body will radiate approximately 648 W/m^2 and the room 419 W/m^2, so you lose 229 W/m^2. That is what you feel as being cooled by the room, from radiation only. Conduction and convection should be added of course. The Earth works the same way—lose some, gain some. It is that balance that is being argued in the whole global warming debate.

KLC: I still feel like I'm missing something. IR heat lamps are pretty efficient, maybe 90%? Let's pick a distance of 1 meter and I want to create a one-square meter flooded with an additional 300W/m^2. It must be additional irradiation, doesn't it? That's going to take a good bunch of lamps and I would feel this heat. However, I go outside and hold out my hand. It's cold. There's no equivalent of 300W/m^2 heater in addition to whatever has heated the ambient air.

Perhaps I'm puzzled by something that is more like a flux—something that just is as a side-effect of a temperature difference and not really something that is capable of doing any work or as a vehicle for transporting heat energy.

It's a canard of climate science that increasing atmospheric CO_2 from 390PPM to 780PPM will raise the Earth's surface temperature by about 1°C (expanded to 3°C by positive feedbacks). From my way of thinking, the only thing CO_2 can do is increase coupling to space—it certainly can't store or trap energy or increase the Earth's peak or 24-hour average temperature.

Any comments are welcome.

MC: Efficiency of a lamp depends on what you want, if heat is what want then they are 100% efficient, because all electrical energy will be converted to heat, the visible light as well, when it is absorbed by the surrounding room. If visible light is required, a light bulb loses a lot of heat compared to an energy saving lamp. Energy cannot be created or destroyed—first law of thermodynamics.

When you say W/m^2 you ARE in fact talking about a flux (heat flow is what will be in W). If you have two objects radiating towards each other, the heat flow direction will be from the hotter one, radiating (emitting) more and absorbing less, to the cooler one, which radiates less and absorbs more (second law of thermodynamics). The amount of radiation emitted from each of them depends on two things ONLY, the temperature of the object and its emissivity. So radiation is not a side effect to temperature, it is THE EFFECT. Anything with a temperature will radiate according to it, and emissivity. (If something is hotter than 500°C we get incandescence, emission of visible light.) Assuming an emissivity of unity, which is what everyone seems to do in this debate, the radiation (flux. integrated from zero to infinity) will be equal to what can be calculated by Stefan-Boltzmann's law, which is temperature in Kelvin, raised to the fourth power, multiplied by that constant sigma. It's that simple!

With regard to your thought experiment, it is always easier to calculate what an object emits than what it absorbs, because emission will be spreading diffusely from an object, so exactly where it ends up is difficult to predict. I am not sure where you are aiming with that idea, but it does not seem to be an easy experiment to do in real life, at least not with limited resources.

CO_2 is a pretty powerful absorber of radiated energy, that fact is well known. Water vapor is an even stronger absorber. In the climate debate it is also considered a reflector, which probably also true, because that is universal. Everything absorbs and reflects to a degree. So I guess that the feedback you mention has to do with the fact that increasing temperature increases the amount of water vapor, which increases absorption, and so on. But my knowledge is pretty much limited to what happens down here on Earth, because that is what matters when we measure temperature using infrared radiation. However, it is important to remember, again, that we talk about different spectral bands, the influx is concentrated around a peak in the visual band and the outgoing flux is around 10 micrometer in the infrared band, and the absorption may not be the same.

With so many scientists arguing about the effects of CO_2 I am not the one to think I have the answers. I really don't know what the truth is. And the problem that all these scientists have is that they will never be able to test if their theories are correct, because the time spans are too long. For a theory to be scientifically proven, it has to be stipulated and tested, and the test must be repeatable and give the same results in successive tests for the theory to be proven.

If not, it is not science, it is guessing.

More like a horoscope…

This interchange stirred a lot of dialog. Often you'll notice in the comments on an article that there is wisdom and intelligent commentary—this is a good example. I will quote the more interesting examples.

Responses and Comments about the *Conversation with an Infrared Radiation Expert*

Murray Duffin: Interesting. Now I am confused at a higher level.

Ian W: When water vapor in the atmosphere condenses into liquid water and then changes state again and becomes ice, it gives off latent heat for both state changes.

Does that latent heat release follow Stefan-Boltzmann's

radiative equation?

Dennis Wingo:

> *With so many scientists arguing about the effects of CO_2 I am not the one to think I have the answers. I really don't know what the truth is. And the problem that all these scientists have is that they will never be able to test if their theories are correct, because the time spans are too long. For a theory to be scientifically proven, it has to be stipulated and tested, and the test must be repeatable and give the same results in successive tests for the theory to be proven.*

Now that is a wise statement. As someone who designs temperature measurement systems (principally for spacecraft) I have always been amused that the AGW community can make the statement that they can use an instrument with 0.5 degree accuracy and get 0.001 degree temperature variation out of it.

In designing thermal control systems for spacecraft you have to be incredibly sensitive to the absorptivity/emissivity (a/e ratio). If you get this even slightly wrong in spacecraft design, you either run the equilibrium temperature too high and it will fail, or too low and it will fail.

MC makes the observation that the radiated temperature is 100% dependent upon emissivity, which is correct, but I have never seen this really integrated into AGW models, they simply use a blackbody approximation, which is a terrible reference in that this varies wildly around the world.

Also, the models do not take into account the dramatic differences in resulting temperature based upon altitude, especially in desert regions of the globe.

A lot of physics modeling operates by making simplifying assumptions, but how many of these assumptions are testable and repeatable?

This is the basis of a lot of my skepticism on the models

involved.

David Ball:

> *The only thing CO_2 can do is increase the coupling to space.*

Thereby having a cooling effect. Good stuff. I would love to hear what Dr. Lindzen has to say about this. If I am not mistaken, this is what he has been saying.

ferd berple: Great comment:

> *For a theory to be scientifically proven, it has to be stipulated and tested, and the test must be repeatable and give the same results in successive tests for the theory to be proven. If not, it is not science, it is guessing.*

Why is it that leading climate scientists have tried to hide their data and methods? In what way are such tests repeatable? In what way have they stood the test of time?

Berényi Péter:

> *MC: For 300 W/m^2 radiation I get -23.4°C at 300 W/m^2 when I calculate it (yes, minus!). Pretty cool light bulb.*

300 W/m^2; is more like -3.3°C (26°F). It is still cold though. At -23.4°C (-10°F) black body radiation is only 220 W/m^2.

chemman: Good post, thanks. The one area I struggle with is the reflector part. Do the "green house gas" molecules absorb and re-radiate energy? Yes. Is it like a reflector? Questionable. Since molecules are 3-dimensional and constantly in motion "green house gases" will re-radiate IR energy in 3 dimensions not just back at the Earth like a reflector. Again it just may be the weakness of the terms used that is throwing me.

dp: The confusion of temperature vs. energy is an interesting one. I've attempted to explain how it works by the following. Place a sheet of paper above a surface. In that sheet is a 6" hole through which passes sunlight. That light lands on a thermometer which records the temperature. It is intuitively obvious the temperature will remain the same if we move the paper and thermometer around the immediate area so we determine there is a fixed relationship between the Sun and the thermometer.

Now we place a 6" lens over the paper and by good fortune the focus point is exactly on the thermometer. We know the same amount of energy is entering the lens as entered the hole in the paper but the temperature is likely to cause the thermometer to explode. The energy is concentrated.

This is clearly not a tale told to people of science with any expectation of creating revelation but it helps the grand kids to understand the relationship between heat and energy.

JEROME: It seems that MC knows the subject matter extremely well. It also seems that he says you cannot be certain what is really going on with IR and CO_2 in the atmosphere, despite those who seem to be absolutely certain.

Food for thought indeed!

Jim D: MC puts KLC right on a few of his misconceptions. I only fault MC's terminology where he uses the word 'reflected' for IR from the sky, when in fact it is emitted. This is probably just loose wording on his part, but there is a scientific distinction.

Charles Nelson: Absolutely brilliant—thank you. I like the term 'increase Earth's coupling with space' I have been trying to explain this to the numpties forever.

One question I ask CO_2 Warmists is, 'have you ever noticed the outside temperature when flying on a plane at cruising altitude:

-30 to -60 degrees C. right? Have you ever looked down and wondered what is happening at the top of the cloud layer—heat loss right? So if the atmosphere was to get warmer wouldn't these clouds simply rise a little higher because of convection and lose their energy through radiation into the vast reservoir of coldness above?

Interestingly I find that when I use words like radiation and convection they tend to glaze over and lose interest. In fact I don't think that 'science' matters that much to believers...

Keep up the good work guys.

Domenic: I am an IR expert, 20+ years in the field.

CO_2 is an IR absorber of only narrow wavelength bands of IR.

Water vapor is a much more stronger IR absorber because it absorbs very large wavelength bands of IR.

In addition, water vapor is approx 3.5% of the atmosphere or 35000PPM. (That's a global average. At the poles the air is drier. In the tropics, the air is much wetter.)
CO_2 is only about 390PPM.

So, water vapor from a PPM point of view is probably 100X greater in effect as a greenhouse gas from an atmospheric percentage only compared to CO_2. (I rounded up slightly because the tropics have more effect than the polar regions, having more water vapor to absorb and store solar energy.)

In addition, a molecule of H_2O is also quite a few multiples greater in absorption of IR compared to a molecule of CO_2.

Take a look at absorption spectra for H_2O: http://webbook.nist.gov/cgi/cbook.cgi?ID=C7732185&Units=SI &Type=IR-SPEC&Index=1#IR-SPEC

For CO_2:

http://webbook.nist.gov/cgi/cbook.cgi?ID=C124389&Units =SI&Type=IR-SPEC&Index=1#IR-SPEC

Just eyeballing these two NIST absorption spectra curves for H_2O and CO_2, it appears H_2O may be at least 10X greater at absorption per molecule than CO_2 is.

Thus, the comparative effects on 'global' warming, changes in H_2O composition in the atmosphere are probably 1000X greater than CO_2 changes in terms of contribution to 'greenhouse effect' give or take a bit for some error =>>>>> approximately 1000 TIMES!!!!

CO_2 is a non-issue compared to water in the atmosphere.

Schadow: In the 1950's, I was working on the Sidewinder air-to-air missile at the Naval Weapons Center at China Lake, CA. Sidewinder's guidance system is built around an infrared sensor. I remember all the "gee whiz" stuff that was observed during the sensor's testing, e.g., it would lock on to a quarter moon in the middle of a hot summer day and even Venus at night. The sensor has undergone much improvement since those days and advanced versions of the missile are still seen under-wing on today's combat aircraft.

Not much to do with this topic but just an interesting side-note on the practical uses of infrared technology.

Domenic: As long as we're on the subject, most people don't realize that the most accurate CO_2 measuring devices also are IR based. They utilize those narrow bandwidths of CO_2 IR absorption to measure the amount of CO_2 in an atmospheric air sample.

A while ago, I took a look at the history of the supposed pristine CO_2 measurements at Mauna Loa. I pulled up two papers (Keeling 1960, and Thoning 1989) describing the methods, calibration protocol followed, etc.

They have been following very good protocol. However, they have to constantly calculate out the effects of the nearby Mauna Loa volcanic activity. Mostly during night times, due to the prevailing winds, the CO_2 measuring devices do jump showing dramatic increases in CO_2 from the volcanoes. They supposedly developed algorithms to eliminate those errors. I haven't looked at those yet. HOWEVER, that activity from Mauna Loa volcano, has another effect that I HAVE NOT SEEN THEM TAKE INTO

CONSIDERATION.

If Mauna Loa volcano is potent enough to send their CO_2 measurements skyward, that means that WARM AIR from the volcano at night, is also affecting their temperature data. It's the same warm air that contains increased CO_2. And I would bet it also has biased their night time temperature data. BUT THERE IS NO MENTION THAT THEY HAVE FACTORED THAT WARM AIR EFFECT FROM MAUNA LOA VOLCANOE OUT OF THEIR TEMPERATURE RECORDS!

In my opinion, the temperature data from Mauna Loa station is greatly suspect from what I can see.

DirkH: Berényi Péter says:

> *300 W/m^2 is more like -3.3°C (26°F). It is still cold though.*
> *At -23.4°C (-10°F) black body radiation is only 220 W/m^2.*

Gives "warmist" a whole new meaning. CO_2 increases might even increase the temp to -2.3 deg C or so.

Steve Reynolds: Ken, I'm glad you are listening to someone that knows what he is talking about. I'll claim to be somewhat of an expert here as well (I design the IR sensors that go into the kinds of instruments you are talking about).

I agree with everything MC said except (as Jim D noted) that IR is not reflected by CO_2, it is absorbed and then re-emitted in all directions.

Mark Wagner:

Since molecules are 3-dimensional and constantly in motion "green house gases" will re-radiate IR energy in 3 dimensions not just back at the Earth.

I doubt that CO_2 re-radiates much at all. As soon as the CO_2 molecule absorbs any radiation, 90% are immediately quenched by contact with other molecules in air, primarily nitrogen. The other molecules don't "radiate" in the sense that CO_2 does, with bending molecular bonds. They just convect and carry the heat up and away. I've never seen any science that directly addresses quenching of "warmed" CO_2 to other atmospheric gasses prior to "re-radiation."

Jim Masterson:

> The remaining 0.9 W/m^2 seems awfully close to what I would assume to be the inaccuracies in the numbers input to calculate it. You are balancing on a very thin knifes edge with such big numbers as inputs for reaching such a small one. An error of $\pm 0.5\%$ on each measurement would potentially throw it off quite a bit, in the worst case.

This is a major problem with Trenberth, Fasullo, and Kiehl 2009. They use the same atmospheric window value of 40 W/m^2 as they do in Kiehl and Trenberth 1997.

There is a minor problem with cloud cover. The cloud cover is supposedly 62%. KT 1997 combines three cloud layers (49%, 6%, & 20%) to get that figure. They call it "random overlap," whatever that means. It looks like an application of the Inclusion-Exclusion principle. I get 61.6% which rounds to 62%. Their famous energy diagram (fig. 7 in KT 1997 and fig. 1 in TFK 2009) should state: "62% cloud cover assumed."

After calculating 62% for the global cloud cover, the term "cloudy" is ambiguous throughout the rest of KT 1997. Every time you see "cloudy," does KT 1997 mean 100% cloudy, 62% cloudy, or something else?

My favorite computation is the value for the atmospheric window, and I quote from KT 1997:

> The estimate of the amount leaving via the atmospheric window is somewhat ad hoc. In the clear sky case, the radiation in the window amounts to 99 W/m^2, while in the cloudy case the amount decreases to 80 W/m^2, showing that

> *there is considerable absorption and re-emission at*
> *wavelengths in the so-called window by clouds. The value*
> *assigned in Fig. 7 of 40 W/m^2 is simply 38% of the clear sky*
> *case, corresponding to the observed cloudiness of about 62%.*
> *This emphasizes that very little radiation is actually*
> *transmitted directly to space as though the atmosphere were*
> *transparent.*

This is really sloppy math. The term "cloudy" is again ambiguous. If KT 1997's cloudy term means 62%, then the correct window value is 80 W/m^2. If they mean 80 W/m^2 is a 100% cloudy value then they should interpolate between 99 W/m^2 and 80 W/m^2 and get something like 87 W/m^2. Apparently the 80 W/m^2 cloudy value is thrown in as a detractor, because they interpolate between 99 W/m^2 and 0 W/m^2. Apparently, they obtain 37.62 W/m^2 and round up to 40 W/m^2.

That's a slop of at least 2.38 W/m^2 (ignoring the other larger values). The 0.9 W/m^2 seems a little nonsensical to me.

Bruckner8: The narrative is so clear, practical, non-confrontational, unemotional and straight-forward that I came away thinking "I *almost* understood what he said." He brought up the measurement accuracy again too, which I posted about in my first post on WUWT a couple years ago, and another commenter has chimed in with his professional experience.

Are there any matter-of-fact scientific narrative examples similar to this on the AGW side?

Steve Reynolds: Domenic, you have a major error in your calculation: you used IR absorption data for 'liquid water', not gas phase water vapor!

Even when you correct that, you need to take into account that absorption reaches near 100% in narrow spectral lines, so adding more does not increase absorption linearly. That is why the effect of CO_2 only increases approximately as the log of concentration.

Tom_R: Jim D says:

> MC puts KLC right on a few of his misconceptions. I only fault
> MC's terminology where he uses the word 'reflected' for IR
> from the sky, when in fact it is emitted. This is probably just
> loose wording on his part, but there is a scientific distinction.

His statement was that there's always reflection. In a sense, that's
true here. Whenever you have a change in the index of refraction
you get a reflection at the boundary. I'm not sure how that applies
to the continuous change as the atmosphere thins with altitude, but
there might be some reflection from that.

G. Karst: Jim D:

> I only fault MC's terminology where he uses the word
> 'reflected' for IR from the sky, when in fact it is emitted.

I believe the reflection, re-emission, back-scattering terms are
describing the same phenomenon, depending on the discipline. In
the field of optics, it is considered reflection. From quantum physics
we have no such phenomenon, as photons are absorbed and re-
emitted. Even in nuclear reactors, old terminology, referred to the
"moderator" as a "reflector". It is a problem during inter-
disciplinary discussions, especially when dealing with photons which
can be regarded as both particle and wave. Personally, I prefer the
term back-scattering as most descriptive of the term/process. GK

Steve Reynolds: Mark Wagner:

> As soon as the CO_2 molecule absorbs any radiation, 90% are
> immediately quenched by contact with other molecules in air,
> primarily nitrogen. ...I've never seen any science that directly

addresses quenching of "warmed" CO_2 to other atmospheric gasses prior to "re-radiation."

The effect you describe is included in the science. Re-radiation is assumed to occur at about the same temperature as the other gases at that altitude. That is why the measurements described in the article give -50 degrees C or thereabouts.

steven mosher: Schadow says:

> *In the 1950's, I was working on the Sidewinder air-to-air missile at the Naval Weapons Center at China Lake, CA. Sidewinder's guidance system is built around an infrared sensor. I remember all the "gee whiz" stuff that was observed during the sensor's testing, e.g., it would lock on to a quarter moon in the middle of a hot summer day and even Venus at night. The sensor has undergone much improvement since those days and advanced versions of the missile are still seen under-wing on today's combat aircraft.*
>
> *Not much to do with this topic but just an interesting side-note on the practical uses of infrared technology.*

Yup, but there is more to it than that. If you want to do IR design today (of sensors, or countermeasures or stealth technology) the physics you use is the same physics that many skeptics deny. Good thing we don't let them design the machines that our safety as a nation depends upon.

W. Falicoff: Domenic—the CO_2 measurements by Scripps Institute are taken in several locations around the world including the locations near the North and South pole.[110]

These measurements support the findings of the measurements at Mauna Loa. They also show there is a lag of CO_2 from the

---

[110] See http://scrippsco2.ucsd.edu/research/atmospheric_co2.html

Northern to Southern Hemisphere (the levels in Antarctica lag those in the Northern hemisphere), as expected given the gains in CO_2 are primarily from anthropogenic sources.

The models used by climate modelers show that the heat balance in the upper atmosphere from increased concentrations of CO_2 is what matters, not its effect at sea level where water vapor dominates .

Dennis Wingo: With regard to your statement that climate modelers assume CO_2 emits as a black body (emissivity of 1) I believe is not true (the emissivity of CO_2 in the IR range is quite low). Further, your statement that it is not possible to achieve a measurement less than the accuracy of an instrument is also not correct, as the accuracy can be increased by taking many measurements (up to a threshold). The resultant accuracy is approximately proportional to the square of the number of readings. This holds true in several fields. For example one can determine the position of a smear image of an object on a digital sensor to the resolution of 1/5 the size of a pixel (so-called super-resolution techniques). The recent findings of Earth-like planets around other stars uses the change in apparent brightness of the star when a planet or plants are in different positions on their orbit. The sensitivity of the systems used in these measurements is extraordinary, well beyond what scientists thought was possible with state-of-art equipment. By the way where did you see that AGW scientists are measuring temperature to an accuracy of one thousandth of a degree (K?)?

tallbloke:
...what can be calculated by Stefan-Boltzmann's law...
The Stefan-Boltzmann Law. Two guys, one law.

Dr. Dan: Ken, you referenced Figure 1 in the Trenberth article. This is very similar to the figure for the energy balance of the Earth in IPCC AR4 FAQ. What strikes me about the energy balance is that about 160 W/m^2 is absorbed by the Earth's surface, of which 17 ascends to the upper atmosphere in thermals and 80 is used to evaporate water. That leaves a net of about 60 W/m^2 to be radiated. Yet, the diagram shows well over 300 W/m^2 being emitted to the CO_2 "cloud". Where does all that energy come from, and what ever happened to the First Law of Thermodynamics that energy can neither be created nor destroyed? There seems to me to be no need of considering the resulting temperature when the energy simply does not add up.

By the way, it matters little that the energy is circulated. That additional 300 W/m^2 has to come from somewhere. It this diagram is correct, then we are getting free energy, in which case we do not need any more power plants at all.

Latitude: Steve Reynolds says:

> *I agree with everything MC said except (as Jim D noted) that IR is not reflected by CO_2, it is absorbed and then re-emitted in all directions*

Steve, since it's constantly being exposed to IR, when it's saturated, does it just reflect because it can't absorb any more? Or is it in a constant state of absorbing and re-emitting and neutralizing itself?

Since it's constantly being exposed to IR, what happens?

John S: Domenic Wrote:

> *Thus, the comparative effects on 'global' warming, changes in H_2O composition in the atmosphere are probably 1000X greater than CO_2 changes in terms of contribution to 'greenhouse effect'."*

This fact has been acknowledged by the AGW community, but as their argument goes, the relatively small added heat from CO_2

creates a positive feedback loop that includes adding more water vapor into the atmosphere until there comes a 'tipping point' where the global temperatures go into thermal runaway.

CO_2 is the "trigger" of this positive feedback AGW, not necessarily the source of all the extra heat.

As a "warmist," I cannot deny that CO_2 does absorb and re-radiate heat energy. It is the existence of positive feedback vs. negative feedback mechanisms in the atmosphere that is at the heart of the global warming debate.

richard verney: What one is measuring is interesting. But one of the problems in this debate is the assumptions made by the AGW proponents and the fact that everything is treated as if it were an average. Averaging can be a useful tool. However, it is almost invariably the case that when one examines any given individual scenario, one is looking at something other than the average. In other words, the average is rarely encountered in the real world experience. This obviously has a bearing when one considers what they wish to see when making a measurement. Ken Coffman ponders upon an experiment where he is hoping to measure back radiation of $300W/m^2$, but this is the so called average back radiation and this raises the question of precisely where on planet Earth would you see that amount of back radiation? It is highly unlikely to be seen at the location where he was conducting the experiment. That being the case, the results of the measuring experiment may tell one little about the effects of CO_2.

I can understand that CO_2 absorbs IR and then re-radiates this in all directions, some up, some sideways and some downwards etc. This scattering is a continuing process such that I can understand that as a consequence of that, CO_2 delays IR finding its way out to space. However, what I do not understand is how this effectively heats up the Earth.

The average global temperature of the Earth will only heat up if the amount of energy received from the Sun is not fully re-emitted back to space (by way of all forms of energy dispersion).

But CO_2 merely delays the escape of IR radiation, it does not provide an impenetrable shield preventing the IR from escaping back to space.

This then begs the question whether the encumbrance/delay caused by CO_2 has any significant impact. The answer to this appears to me to be whether the delay is of such magnitude that all the IR does not have time to escape to space during the time when the Earth is not receiving energy from the Sun. Whilst this process is happening 24 hours a day, simplistically the question is whether during the period of night there is sufficient time for all the energy received during the day to escape back into space. If there is sufficient time during night for this to happen, then there is no effective energy entrapment and the Earth will not be heating up.

IR travels at the speed of light. Approximately 300,000 km per second. If a photon of energy was to take a direct route to space this would involve a distance of about 120km. If because of the presence of CO_2 in the atmosphere, the photon does not take a direct route but instead bounces around millions of times taking a zigzag course into space (bearing in mind that the density of CO_2 decreases with height), we are delaying the IR escape into space by minutes. It is clear that all the energy received during the day, can (subject to water vapour and clouds etc) escape into space within less than an hour of sunset even if the concentration of CO_2 were to double. So how does CO_2 cause warming in the real world in which we live in?

John Cooper: I had to dig deeply through my old space program files to find this one:

> *I fully realize that I have not succeeded in answering all your questions. Indeed, I feel I have not answered any of them completely. The answers I have found only serve to raise a whole new set of questions, which only lead to more problems, some of which we weren't even aware were problems. To sum it all up, in some ways I feel we are more confused than ever, but*

I believe we are confused on a higher level, and about more important things.

Bill Illis: Really nice article.

I would like to see much more discussion of the actual physics involved here because this is all happening at the quantum level—where physics is king—not in climate models where 20 km-square boxes and just 21 layers of the atmosphere is king.

The numbers in this debate are staggeringly huge as well as staggeringly small.

The energy represented by a solar photon spends an average 43 hours in the Earth system before it is lost to space. Some spend just a millisecond while a very, very tiny percentage might get absorbed in the deep ocean and spend a thousand years on Earth or longer. In essence, the Earth has accumulated 1.9 days worth of solar energy. If the Sun did not come up tomorrow, it would take around 86 hours for at least the land temperature to fall below - 200°C.

The energy represented by a solar photon spends time in 5 billion individual molecules on Earth before it escapes to space. That means it is bouncing around from molecule to molecule to molecule almost continuously. The IR emitted by the surface is not skipping Nitrogen and Oxygen molecules and preferentially seeking out CO_2 and H_2O only. Every molecule on Earth and in the atmosphere is participating in this process and does so continuously. Maybe CO_2 or H_2O provides the initial absorption, but that energy is shared amongst the rest of the atmospheric molecules almost immediately. What happens to it then?

The surface accumulates almost none of the solar energy which hits the surface during the height of the day. 960.000 joules/m^2/second is coming in and 959.083 joules/m^2/second is moving up and away from the surface. At night, virtually no energy is coming in and only 0.001 joules/m^2/second is flowing up and out to space. That is not consistent at all with the greenhouse theory and the back-radiation theory. It is more consistent with energy flowing from hot to cold continuously (like the second law of thermodynamics) and it flows faster the more there is a differential between that hot and cold.

We need a time perspective on radiation physics because it is happening at the speed of light and at the miniscule amount of time that a molecule absorbs that energy before passing it on through emission or collisional exchange. CO_2 holds onto an absorbed IR photons for an average 0.000005 seconds before it is emitted or passed onto another molecule, Every atmospheric molecule hits another atmospheric molecule every 0.00000000015 seconds, an emitted IR photon from the surface could escape the atmosphere in just 0.000016 seconds at the speed of light—yet it actually takes 40 hours to make the journey. In the Sun, the average photon takes 200,000 years to make it out.

Is there a climate model that can simulate that accurately? It would be far too complicated for any computer even 1000 years from now to be able to simulate accurately. We have to empirically measure what is really happening in such a complicated system and base models on that instead. What is really happening is that the theory is off by at least half to date.

Hans Erren: Adding water vapour in the tropics gives the tropics a greenhouse effect, I remember very well when I left the air conditioned hotel in Dhaka Bangladesh, a wall of humid heat hit me and I had to wipe my glasses.

Every added particle to the atmosphere is a small downward radiator, adding more particles therefore adds more radiators. In a nutshell, that's the greenhouse effect.

u.k.(us): steven mosher says:

> *Yup, but there is more to it than that. If you want to do IR design today (of sensors, or counter measures or stealth technology) the physics you use is the same physics that many skeptics deny. good thing we don't let them design the*

machines that our safety as a nation depends upon.

WOW, did you just say skeptics are too stupid to defend themselves??
Sounds like you helped protect us, thanks for that!!
I'll say it again, thanks.
Now, it is time to protect ourselves, from ourselves.

wayne: Mark Wagner says:

> *I've never seen any science that directly addresses quenching of "warmed" CO_2 to other atmospheric gasses prior to "re-radiation."*

Mark, this has been discussed here a couple of times before.[111] [112]
If anyone else wants something to consider: this is a two way street and the energy moving from CO_2 to N_2 and O_2 can go backwards to both H_2O and CO_2 eventually, for all energy does leave the Earth only by radiation. And since there are some 100X more H_2O molecules up to the Tropopause, mostly H_2O accepts this energy and radiates it to space.
[And yes, a bit less than half is directed downward, but "back radiation" is a null effect when speaking of surface to space transfer, it can't actually re-warm the surface in bulk (thermodynamically), only if you speak of singular photons and ignore the other effect (cooling) that cancels.]
If you are speaking of published scientific papers I have not seen any either.
All of Mikael Cronholm's explanations are well worded and even the mention of reflection can be quite correct if placed in the right frame of reference. All reflection is an absorption and re-

[111] http://wattsupwiththat.com/2010/08/05/co2-heats-the-atmosphere-a-counter-view/
[112] http://wattsupwiththat.com/2010/08/31/does-co%e2%82%82-heat-the-troposphere/

emission at the quantum level for Richard Feynman made that so perfectly clear in his explanation of his quantum electrodynamics.[113]

If you listen to his lectures, just replace the glass with the atmosphere and the light with infrared and ask: Can the IR cancel just as he describes black reflection bands as his mental counter goes Zzzzzzzz..... (on the complex plane)?

Richard Bell: Wonderful stuff—I would like to confirm that CLOUDS (shouting loudly!) are the winners in my book. I have just spent 5 days on a very remote beach on the west coast of Baja, no people, no lights, nothing, just the big blue watery thing called the PACIFIC, the big yellow thing called the SUN and on one cold windy day some white stuff called CLOUDS—they had a stunning effect for that day, then were gone.

In the isolation that I was experiencing the big three just made sense—thanks for the very interesting IR info and discussion.

DirkH: John S says:

> ... This fact has been acknowledged by the AGW community, but as their argument goes, the relatively small added heat from CO_2 creates a positive feedback loop that includes adding more water vapor into the atmosphere until there comes a 'tipping point' where the global temperatures go into thermal runaway.

The tipping point, if it exists, should be observable first in hot, moist regions, then. So places like the Amazon should "tip" before, say, sub-zero wintery Siberia. Once a hot, moist place on Earth "tips over," the generated increased moisture should spread, making neighbouring regions "tip over" into the "hothouse Earth" mode.

In fact, the threshold difference between the Amazon rain forest and Siberia should be so big that a much smaller CO_2

---

[113] http://vega.org.uk/video/subseries/8

concentration should suffice to make the rainforest tip over initially than Siberia.

So, if we want to find proof for the existence of the tipping point, we should watch the rainforest. And maybe one could do an experiment. CO_2 is heavier than air. Erect a big circular "fence" of thin plastic, maybe lifted up with a number of balloons, and increase the CO_2 concentration within the fence. That should bring us much closer to the tipping point. As surface LWIR is absorbed and re-emitted by CO_2 after a few meters, the effect should be measurable even if the fence is only say a 100 m high.

The fence would also ensure that the "tipping" can't spread to pristine wilderness. Any takers?

Horace the Grump:

> For a theory to be scientifically proven, it has to be stipulated and tested, and the test must be repeatable and give the same results in successive tests for the theory to be proven.
>
> If not, it is not science, it is guessing.
> More like a horoscope...

Completely nailed AGW to the mast—or he could have said "More like a religion..."

Either is probably acceptable...

Molon Labe: steven mosher says:

> the physics you use is the same physics that many skeptics deny

"Many skeptics," indeed. Well, "some say" they are sick and tired of strawman characterizations of their skeptical positions.

Ian W: John S says:Domenic Wrote:

> Thus, the comparative effects on 'global' warming, changes in

H_2O composition in the atmosphere are probably 1000X greater than CO_2 changes in terms of contribution to 'greenhouse effect'.

This fact has been acknowledged by the AGW community, but as their argument goes, the relatively small added heat from CO_2 creates a positive feedback loop that includes adding more water vapor into the atmosphere until there comes a 'tipping point' where the global temperatures go into thermal runaway.

CO_2 is the "trigger" of this positive feedback AGW, not necessarily the source of all the extra heat.

As a "warmist," I cannot deny that CO_2 does absorb and re-radiate heat energy. It is the existence of positive feedback vs. negative feedback mechanisms in the atmosphere that is at the heart of the global warming debate.

It is more the insistence that there is one linear positive feedback—the rate of water evaporation into the atmosphere and then water vapor acting solely as a 'green house gas' that is at the heart of the debate. For example, if the water vapor feedback is initially positive but then becomes strongly negative due to albedo effects of clouds and release of latent heat at height above the optically thick CO_2/H_2O layers of the atmosphere (this appears to be shown to be the case by satellite metrics) then the AGW hypothesis fails.

Jim D: I will clarify my earlier statement regarding IR emission versus reflection. Reflection of IR photons is negligible if you remember (e.g. from "why is the sky blue?" arguments) that scattering is greater for shorter wavelengths (scattering being reflection), and IR has a long wavelength relative to visible light. The measured IR from the sky is pretty much all emitted by CO_2 and H_2O molecules (and the other greenhouse gases present). You don't see the same photons coming back down as the ones going up from the Earth.

hotrod (Larry L): Good discussion that adds to the recent posts I made about experiments using an IR thermometer to get a sense of the "effective" temperature of the sky as seen from the ground.

The important thing that this experiment shows, is that unless the ground or ocean surface is receiving direct sunlight its entire view to the sky is in effect a very, very cold surface. As you mention it is frequently off scale low on my IR thermometer which reads down to -76 deg F (-60 deg C) at emissivity 0.95 setting.

As a result a surface like the ground, a roof, or the ocean's surface **must always** be losing heat to the sky by IR if it is not directly illuminated by the Sun (or scattered sunlight from clouds etc.) The only other source of heat energy to maintain the temperature of that surface must come from heat transferred from the air, which by comparison is very small unless there is air motion.

It is perhaps a subtle distinction for some folks, but the CO_2 and water vapor "back radiation" does not "warm the ground" or the ocean's surface, it only "reduces the rate at which it is cooling by IR radiation" to the apparent surface of the very, very cold sky.

As I showed in the measurements I made in the IR, due to this constant IR exchange between surfaces and the cold sky, there is very little relationship between the air temperature and nearby surfaces. At night, concrete is a very efficient radiator in the IR being like asphalt a nearly perfect black body radiator with an emissivity in the high 90% range (different sources give different values but most cluster between 0.95-0.98 for both).

What is more important, in a built up area, is a given piece of ground has a significantly reduced view of the sky due to buildings. As a result it cannot cool as rapidly through IR radiation to the sky at night because most of its hemispherical view is no longer of the open sky but rather building walls that are only a few degrees cooler than the air temperature rather than deep into the subzero IR temperatures of the open sky.

The urban heat island effect is as much to do with changes in visibility of the sky, as it is about waste heat lost from buildings.

Even if you built an empty town that used no energy, ground measurements would still be warmer at night due to this change in the IR view of the sky due to structures and modifications of the topography that limit the hemisphere of the sky that is viewable from any given surface.

An analogy could be made if you replaced IR radiation with visible light. The sky would be a deep black but all nearby surfaces are very bright. The illumination of the ground surface would be completely dominated by the bright surfaces in a built up area, where in the open prairie or desert the illumination of a flat surface would be completely dominated by the black sky, with only a trace of scattered light from the nearby ground.

No matter how much you insulate buildings or reduce their energy consumptions, you will not change this IR view issue to the sky. Surfaces will always be warmer in a built up area than they would be in the open unobstructed environment of open ground regardless of energy consumption.

Richard Sharpe: Hans Erren says:

> *Adding water vapour in the tropics gives the tropics a greenhouse effect, I remember very well when I left the air conditioned hotel in Dhaka Bangladesh, a wall of humid heat hit me and I had to wipe my glasses.*
>
> *Every added particle to the atmosphere is a small downward radiator, adding more particles therefore adds more radiators. That's in a nutshell the greenhouse effect.*

Hmmm, surely, it is every second particle?

Secondly, which particles? H_2O, CO_2, O_2, N_2?

Which has the larger effect?

Steve Reynolds: Latitude:

> Since it's constantly being exposed to IR, when it's saturated,
> does it just reflect because it can't absorb any more?

There is no reflection from a gas. Also, under normal conditions, CO_2 in the atmosphere does not become 'saturated' if I understand how the term is being used.
　　Latitude:

> Or is it in a constant state of absorbing and re-emitting?
> neutralizing itself?

Yes, molecules are absorbing, equilibrating temperature with surrounding molecules, and emitting at the equilibrated temperature in random directions.

Latitude: Steve Reynolds says:

> Yes, molecules are absorbing, equilibrating temperature with
> surrounding molecules, and emitting at the equilibrated
> temperature in random directions.

So it's just absorbing and scattering the IR, same in and same out. Wasn't clear on that, I slept through class…
　　Thanks, Steve

Dave Springer says:

> …bulbs emitting $300W/m^2$ is simply impossible, because
> $300W/m^2$ corresponds to a very low temperature!

It wouldn't be a conventional "bulb" first of all because those are nowhere near blackbody spectrum and second of all because they're near infrared not far infrared.

Your "light bulb" would be a piece of black metal with an area of 1 square meter maintained at a temperature of -23.4°C. In most cases you'd have to be cooling it below ambient temperature unless you were operating it outdoors the other night in Oklahoma when it hit -35°C then you'd need to run a bit of electricity through it to heat it up some.

This is essentially what the great experimental physicist John Tyndall did 150 years ago when he experimentally verified infrared opacity in various gases. Tyndall's main "light bulb" was a copper vessel painted with lamp-black filled with boiling water which emits a blackbody spectrum in the far infrared and maintains a pretty constant temperature without much hassle since water won't heat above its boiling point without pressurizing the vessel. It was aimed at a brass tube polished to a mirror finish on the inside and capped on each end with a plate of rock salt (rock salt is practically transparent to infrared) with valves and a vacuum pump to evacuate the tube and introduce the gases he wanted to test. On the far end had a thermopile and galvanometer.

The galvanometer response was non-linear and so to keep the thermopile outputting a voltage in the linear high-sensitivity range of the device he put another "infrared source on the backside of the thermopile with an adjustable shield to regulate how much radiation the thermopile received from the backside. The setup was quite sensitive and just a warm body in the same room with it would compromise the experiments. To demonstrate its sensitivity Tyndall would bring the thermopile/galvanometer into halls where he was giving a lecture, aim it at a wall on the other side of the lecture hall then have a member of the audience step into its view and the galvanometer would peg at the max reading. So a warm body moving around the lab during an experimental runs didn't compromise the results he read the galvanometer from a distance with a telescope. It was quite the ingenious experimental setup given the times (1850).

The thermopile was quite sensitive and just a warm body in the same room with it threw it off.

DirkH: Bill Illis says:

> The IR emitted by the surface is not skipping Nitrogen and
> Oxygen molecules and preferentially seeking out CO_2 and H_2O
> only.

I would say the IR photons *do* skip the N_2 and O_2 molecules as these
molecules can't absorb IR. The IR photon does not "see" these
molecules but passes through them as if they weren't there.

Of course, after the IR photon is absorbed by, say CO_2, the
energy is thermalized and can be transferred from, say CO_2 to N_2—
but also back from N_2 to CO_2, and the CO_2 can re-emit an IR
photon.

eadler: Charles Nelson says:

> Absolutely brilliant—thank you. I like the term 'increase
> Earth's coupling with space' I have been trying to explain this
> to the numpties forever.
>
> One question I ask CO_2 Warmists is, 'have you ever
> noticed the outside temperature when flying on a plane at
> cruising altitude...-30 to -60 degrees C. right? Have you
> ever looked down and wondered what is happening at the top
> of the cloud layer—heat loss right? So if the atmosphere was
> to get warmer, wouldn't these clouds simply rise a little higher
> because of convection and lose their energy through radiation
> into the vast reservoir of coldness above?'
>
> Interestingly I find that when I use words like radiation
> and convection they tend to glaze over and lose interest. In
> fact I don't think that 'science' matters that much to
> believers...
>
> Keep up the good work guys.

Brilliant. Focus on one particular phenomenon, radiation from the
tops of clouds, as if that is all that is going on. No need to look at

any other effect to determine the plausibility of the idea that GHG's emission are going to warm the planet. No need to do any mathematical modeling. All you need to do is believe what you want to believe, and select those phenomena which confirm your beliefs and focus on it.

It is great propaganda and apologetics, but hardly science.

AW:

> Brilliant. Focus on one particular phenomenon, radiation from the tops of clouds, as if that is all that is going on.

Well said, we should stop focusing on CO_2, as if that is all that is going on.

Dave Springer: There's no need to be confused. The "canard" the author refers is no canard but a manifestation of his ignorance. The infrared absorption characteristics of various gases was measured experimentally 150 years by John Tyndall. If you shine far infrared light through a column of IR-transparent gas the amount of radiation you get on the far side of the column is the same as you get going through a column of vacuum.

If you replace the transparent gas (like nitrogen) with an IR-absorptive gas (like water vapor) there is less radiation emitted at the far end of the column. The gas absorbs the radiation from a directional source and re-emits it in all directions.

In Tyndall's apparatus there were only two exits for the radiation since the column was a brass tube polished to a mirror finish on the inside—the two exits were plates of rock salt at either end of the tube. The radiation entering the column remains the same no matter what gas (or vacuum) is in the column but the radiation exiting the column at the far end varies according the kind of gas in the column. Energy must be conserved so what's missing at the far end of the tube has to be going somewhere—in fact it heats the gas which increases the amount of energy the gas is emitting and since the gas emits in all directions some of the energy goes through the rock salt entrance plate instead of the exit plate while some smaller portion will heat the brass tube through conduction and some even smaller amount will heat the tube through absorption because even brass polished to a mirror finish isn't quite 100% reflective.

LazyTeenager:

For 300 W/m^2 radiation I get -23.4°C at 300 W/m^2 when I calculate it (yes, minus!). Pretty cool.

I get the impression that there is some miscomprehension here.

I would say that the flux of 300W/m^2 being calculated here is not the same as the 300W/m^2 of down-radiated atmospheric IR.

Maybe the definition of terms and conditions in the SB formula needs to be verified.

wayne: Hans Erren says:

Adding water vapour in the tropics gives the tropics a greenhouse effect, I remember very well when I left the airconditioned hotel in Dhaka Bangladesh, a wall of humid heat hit me and I had to wipe my glasses.

*Every added particle to the atmosphere is a small **downward** radiator, adding more particles therefore adds more radiators. That's in a nutshell the greenhouse effect.*

That's right. Every added particle to the atmosphere is a small upward radiator also—adding more particles therefore adds more radiators. That in a nutshell is why there is no real "greenhouse effect." The two effects always cancel. You say 'more warming downward' and I say 'also equal cooling upward'.

You see, it's a two way street in real physics, you just ignore the part that doesn't fit your (incorrect) view.

That to me that is why Miskolczi came up with the results he did. Now why, no one has answered that. Is it that with the small increase of temperature we have seen and the small expansion of the atmosphere does cause a higher fraction to escape to space because we do live on a sphere and not an infinite plane and that is merely a geometric effect (ex-secant correction)? Could it be that there is one or more negative feedbacks? Is it a combination of both? I'm still searching.

Domenic: To Steve Reynolds: There is not a great deal of difference, if any, between liquid phase H_2O spectral transmission and water vapor spectral transmission. I used those two graphs because data from NIST tends to be more reliable than most that I have seen.

But here's another graph for you.[114]

See the wide absorption band at 5.5 to 7.7 microns—that is entirely by water vapor, H_2O. Compare it to the narrow bands absorbed by CO_2 in the same graph. In the vapor state, it is still at least 10X greater for this wavelength spread.

There is a problem though, and that is that we have to speculate over the entire spectral output of the dominating forces especially the Sun, which outputs much energy in shorter wavelengths to the Earth. To my knowledge, that data (spectral absorption and transmission and reflection of CO_2 and H_2O) has never really been assembled as there had never seemed to be a need for it.

CO_2 does indeed have a reflective component as does H_2O. Only a black body has no reflective component. The reflective component of H_2O is better known than that of CO_2 because water is more common, and has been tested for IR more thoroughly. All real world materials are reflective, absorptive and emissive. In general, metals are highly reflective (think Al foil, etc) on one end of the scale, organics and others less so, water being one of the

---

[114] http://en.wikipedia.org/wiki/File:Atmosfaerisk_spredning.gif

most absorptive, least reflective. And the reflective component is highly wavelength measured dependent, as well as angularly dependent (the angle with which the wavelength strikes the molecule).

But for magnitudes of the effect on the atmosphere in terms of greenhouse effect, the known data is sufficient to see the HUGE differentials between water vapor effects vs. CO_2 effects.

eadler: Dennis Wingo says:

> With so many scientists arguing about the effects of CO_2 I am not the one to think I have the answers. I really don't know what the truth is. And the problem that all these scientists have is that they will never be able to test if their theories are correct, because the time spans are too long. For a theory to be scientifically proven, it has to be stipulated and tested, and the test must be repeatable and give the same results in successive tests for the theory to be proven. There is no such thing as absolute proof of a theory in science. We don't need that in order to take action on the basis of the knowledge that we have.
>
> More detailed observations can tell us more precisely what is happening in the present, so that we don't have to wait 100 years to find out the outcome of increasing CO_2 concentrations. That is the purpose of the scientific research that is going on. Observations of radiation, behavior of clouds, and other phenomena that are involved, plus innovation in computers and models, provide the information needed to produce better predictions of climate change.

Now that is a wise statement. As someone who designs temperature measurement systems (principally for spacecraft) I have always been amused that the AGW community can make the statement that they can use an instrument with 0.5 degree accuracy and get 0.001 degree temperature variation out of it.

No one is claiming that as far as I have seen.

In designing thermal control systems for spacecraft you have to be incredibly sensitive to the absorptivity/emissivity (a/e ratio). If you get this even slightly wrong in spacecraft design, you either run the equilibrium temperature too high and it will fail, or too low and it will fail.

MC makes the observation that the radiated temperature is 100% dependent upon emissivity, which is correct, but I have never seen this really integrated into AGW models, they simply use a blackbody approximation, which is a terrible reference in that this varies wildly around the world.

I think you are wrong about this. Models provide for different emissivities for snow, forests, grassland, rocks, ocean etc.

Also, the models do not take into account the dramatic differences in resulting temperature based upon altitude, especially in desert regions of the globe.

This is clearly wrong. Models do take altitude into account in the prediction of climate change.

See: Elevation Dependency of the Surface Climate Change Signal: A Model Study[115]

A lot of physics modeling operates by making simplifying assumptions, but how many of these assumptions are testable and repeatable?

These assumptions are constantly being tested. One of the first assumptions made, by Arrhenius in 1896, in his modeling of CO_2's impact on global temperature was that average Relative Humidity would be constant and water vapor concentration in the atmosphere would increase, resulting in positive feedback for temperature increase. Satellite observations are being made which seem to

---

[115] http://journals.ametsoc.org/doi/abs/10.1175/1520-0442%281997%29010%3C0288%3AEDOTSC%3E2.0.CO%3B2

confirm this assumption.[116]

In spite of significant biases in tropospheric temperature and humidity in climate models [John and Soden, 2007] and resultant compensating effects in simulating the clear-sky OLR, our analysis finds broad consistency between the observed and modeled rates of clear-sky OLR radiative damping. This consistency is noted over a broad range of observable sources of climate variations, suggesting that the strong correlations between water vapor and temperature necessary to generate such sensitivities are a robust feature of both models and observations. This analysis offers further evidence to support the ability of climate models to depict the physical processes related to the combined water vapor and temperature climate feedback.

This is the basis of a lot of my skepticism on the models involved.

It seems then that the basis of your skepticism is not real.

Dave Springer: Domenic says:

> In addition, water vapor is approx 3.5% of the atmosphere or 35000PPM. (That's a global average. At the poles the air is drier. In the tropics, the air is much wetter.)

True enough. It varies between 1% (Sahara) and 5% (Amazon) except for Antarctic interior where it is close to 0%. But that's at the surface. Adiabatic lapse rate squeezes out the water vapor with falling temperature while CO_2 concentration remains constant with altitude.

Tyndall (circa 1850) however found that the concentration of the gas doesn't matter but rather the total amount of the gas in the column is all that matters. He confirmed that by varying the

---

116

http://news.cisc.gmu.edu/doc/publications/Chung%20et%20al%2020 10.pdf pressure and length of the column. Tyndall performed literally thousands of experiments with varying, varying pressures, varying column lengths, and varying infrared frequencies.

To begin understanding the physics of greenhouse gases one must at least be familiar with what was experimentally demonstrated by physicists in the mid-19th century.

There's plenty to complicate the situation in the real world beyond comprehension because the atmosphere is a dynamic system with varying gases, pressures, and many physical processes other than radiation going on that are moving heat from one place to another and radically different and rapidly variable rates.

Dennis Wingo:

> ...the physics you use is the same physics that many skeptics deny.

This statement is an insult to the intelligence of people who actually work in the field. The parameters related to the absorption and re-emission of IR radiation by CO_2 were worked out by the USAF (the early parts were classified), in what used to be called "upper atmospheric research". Spectrometer technology was specifically improved to the point to where the individual quantum absorption and emission lines were discerned by the mid-late 1950's.

These measurements were used as a validation of the Gaussian to Lorentz transform that characterizes the increase in the absorption/emission lines of CO_2 and other IR absorbers/emitters. This was used to design the IR sensors of ALL of our inventory of IR sensing missiles.

If you find the equations for this emission/absorption, you will find that the increase in the line widths is proportional to the increase in the minor gas against the ENTIRE atmosphere, not just the concentration of the gas relative to the arbitrary baseline of 280PPM. The absorption/emission of CO_2 has two dependent variables, temperature and pressure, neither of which are accurately characterized in the models used by the AGW community.

All of this is set forth in any Quantum Mechanics book on the theory of light, of which the relevant IR wavelengths absorbed/emitted by CO_2 are a part. The relevant text I use is from Loudon, pages 82-89.

Why don't you have the moral courage to look that up and derive the effect of the increase of CO_2 yourself. You will be surprised at the result.

Policyguy: Anthony, this exchange was a treat, it amazes me sometimes what can be learned on the web.

This quote...

> For a theory to be scientifically proven, it has to be stipulated and tested, and the test must be repeatable and give the same results in successive tests for the theory to be proven.
> If not, it is not science, it is guessing.
> More like a horoscope...

...Is intriguing as well. Gore and company stopped long ago trying to prove anything. They were more interested in having enough of their theories "accepted" to stay at the grant trough. In a way it is reminiscent of the times when it was "accepted" that the Sun orbited the Earth. But then it was more for religious reasons than scientific. Today Gore strives to maintain his religious acceptance of CAGW to prop up the price of his offset credits.

J. Bob: W. Falicoff says:

> Further, your statement that it is not possible to achieve a measurement less than the accuracy of an instrument is also not correct, as the accuracy can be increased by taking many measurements (up to a threshold). The resultant accuracy is approximately proportional to the square of the number of readings.

Hmmmm. If a basic sensor has a mean error of 0.75 deg., and 3 sigma of say ±0.25 deg., your true error will be between 0.5 & 1.0 deg no matter how many readings you take. You might want to add a few qualifiers.

Dave Springer: A safe statement is "increasing atmospheric CO_2 will result in increasing surface temperature if nothing else changed." What causes "confusion at a higher level" is that lots of other things DO change and the amount of warming directly attributable by CO_2 infrared absorption and re-emission is small and can be utterly swamped (lost in the noise) by many other dynamic processes.

Water in all its phases has the starring role in shaping our climate. CO_2 plays a major role in surface temperature only during so-called "snowball Earth" episodes when most of the water vapor has been frozen out of the atmosphere and there's little to no liquid water presenting on the surface. In that case there are no working CO_2 sinks so volcanoes, which keep on belching out CO_2, gradually build up the amount of CO_2 in the column until there's enough greenhouse effect to begin melting the planet. If it weren't for CO_2 the Earth would likely be covered in ice with no hope of ever melting. That's the most commonly accepted hypothesis at any rate and I personally haven't seen anything that seriously disputes it.

_**Jim:** Maybe those who doubt the 'frequency selective' nature of CO_2 IR spectra should investigate Molecular Spectroscopy as it relates to the vibrational modes of CO_2 (and the other important GH gas WV) on account of its molecular properties:

> By examining the emission spectrum of the CO_2 laser, we are able to understand much about the CO_2 molecule and about the dynamics of diatomic and triatomic molecules in general. The CO_2 laser is a molecular laser, meaning that it generates light from the vibrations and rotations of the CO_2 molecules in

the plasma rather than from electronic transitions between energy levels, as in a He-Ne laser.

Like a spring between two masses, the binding forces between the atoms of the CO_2 molecule cause the atoms to move in one of three vibrational modes: the symmetric stretching mode, asymmetric stretching mode, and the bending mode.[117]

High-school level physics, boys...

Richard Sharpe: Hans Erren says:

> *Adding water vapour in the tropics gives the tropics a greenhouse effect, I remember very well when I left the airconditioned hotel in Dhaka Bangladesh, a wall of humid heat hit me and I had to wipe my glasses.*
>
> *Every added particle to the atmosphere is a small downward radiator, adding more particles therefore adds more radiators. That's in a nutshell the greenhouse effect.*

wayne says:

> *That's right. Every added particle to the atmosphere is a small upward radiator also; adding more particles therefore adds more radiators. That's in a nutshell is why there is no real "greenhouse effect". The two effects always cancel. You say 'more warming downward' and I say 'also equal cooling upward'.*

What you have said, Wayne, does not seem correct to me.

It seems to me that approximately half the energy absorbed by CO_2 would be re-radiated upwards and half downwards.

They do not cancel.

The energy remains in the atmosphere for a little longer than it

---

[117] *http://en.wikipedia.org/wiki/Infrared_spectroscopy*

otherwise would, but I suspect that it does not matter as H_2O transports much more energy out of the atmosphere and any that is "trapped" by CO_2 quickly departs during the night.

G. Karst: There is no reason for the warmist to exhibit glee to the admission of CO_2 narrow back-scatter effect. All substances absorb and emit IR (yes, even salt and quartz windows). It is all about spectrum or frequency and a molecule's representative target cross sectional area.

The debate is entirely about CO_2 back-scatter's significance on the actual climate.

Since theoretically, CO_2 IR spectrum is already saturated, any additional molecules, simply cannot affect much. The same is not true concerning the planets biomass. It will expand exponentially to increased temps, increased CO_2, and increased available moisture. This too will have a feedback. As stated many times, this is about climate sensitivity as per much debated feedbacks and forcings.

wayne: Dennis Wingo says:

> The absorption/emission of CO_2 has two dependent variables, temperature and pressure, neither of which are accurately characterized in the models used by the AGW community.

Dennis, I see your great point but could you expand a bit on your statement above? I'm wondering if that could be temperature and density, for a slightly warmer atmosphere will expand but the mass remains constant, therefore, density drops but the pressure remains constant at sea level (average that is). Does it make sense?

kuhnkat: Anthony, doesn't your camera "see" in the near infrared? The Earth emits in the far infrared. The cut-off is at about 4μm between the two I believe.

Some of that wonderful military hardware some people keep talking about is working in the near infrared and has little applicability to the back radiation issue!

The scatter, reflection, absorption bit I have been trying to clarify also. Reflection generally refers to the effect where a particle or wave encounters and leaves an object or field at the same angle. Scatter is where the direction is random. Absorption, of course, is where there is no exiting wave or particle. They are different.

Electromagnetic radiation in the atmosphere is rarely reflected by gasses, only particles like water droplets or aerosols. They are either scattered or absorbed.

Someone please correct me as I am obviously not the expert here.

Domenic: To Dave Springer, re the measurement of IR absorption.

The polished tube method is not exact. It's an approximation. It does not account for angular effects of wavelengths and reflective effects of wavelengths striking the molecules accurately. In addition, it relies on imperfect 'spectral window transmission coatings' on the sensor window material, and other factors.

The science is nowhere as accurate as many believe it to be.

chico sajovic: "Back Radiation" is such a horrible analogy, description or way to think about heat transfer in the atmosphere, that I wish it would just go away. What we are concerned with is the flow of heat from the surface of the Earth to space. It's not that the surface warms the CO_2 in the air then the CO_2 "back radiates" re-warming the surface, it's that when the CO_2 in the air is warmed the flow of heat from the surface to the warmed CO_2 is reduced, thus increasing the temperature of the surface until the flow of heat returns to equilibrium.

- Sun rays heat the surface of Earth
- the warmed surface radiates, heating greenhouse gasses
- smaller temp delta between surface and greenhouse gasses reduces heat flow

- reduced heat flow from surface increases surface temperature
- equilibrium is restored with higher temp surface and greenhouse gas

My problem with the greenhouse effect is: how do you separate radiation from the effects of evaporation, condensation, conduction and convection? Anecdotally I feel that conduction is a far superior mode of heat transfer and evaporation is even better.

To cool my cup of hot chocolate faster I should:

a) hold it up to the sky on a low humidity night.

or

b) blow on it.

When I am hot I should:

a) stand outside naked on a low humidity night.

or

b) get soaked in water and stand in front of a fan.

Fitzy: Brilliant conversation, thanks for sharing.

Considering the energy smart rattle being waved by local authorities, one wonders how derelict Urban Authority, IR emission equipment calibration, will be.

And I've learnt the hard way that the moment one puts Red on a map, some policy wonk will smell promotion and run with even the most tenuous data, and as this site so often points out Policy = Misspent funding. Usually when one says IR, the colour red shows up somewhere, and RED = Hot right and away we go.

One may point out a scale can be quite tight, with the banding representing a single degree, case in point, IR work done in winter. All of a sudden ones city glows bright red, the Policy wonk will scream 'Look at all that energy wasted!', until Johnny lab coat points out, 'That sir is a mere 14 degrees centigrade'. And you'd be amazed when you point out that water holds a lot of heat, their eyes cross, they don't like the notion the very landscape is bleeding heat.

I'd be keen to know what tolerances and ranges equate to a fair IR reading within an urban environment, and with emissivity being so key; one can imagine Urban Authorities taxing by the W/m radiated.

Good luck with your endeavors.

kuhnkat: Oh, no, the Ham and radio guys are gonna get me.

In my statement above I left out the effects of IONIZED gasses which DO reflect electromagnetic radiation.

Dave Springer: Steve Reynolds says:

> *Even when you correct that, you need to take into account that absorption reaches near 100% in narrow spectral lines, so adding more does not increase absorption linearly. That is why the effect of CO_2 only increases approximately as the log of concentration.*

Actually that's not quite right. At low CO_2 concentrations it's nearly linear and as concentration rises it becomes logarithmic. John Tyndall experimentally discovered that too by experiment over 150 years ago. It's pretty much a case of the proverbial low-hanging fruit. The first few molecules get all the fruit they can handle but as the number of molecules increase there's less and less fruit available per molecule. IIRC correctly shows the curve is logarithmic by the time concentration reaches 100PPM.

Mikael Cronholm: WOW! When Ken asked me if he could have our conversation offered for publication somewhere I had no idea it would create such an interest and so many extremely initiated comments. I am glad I could contribute even if it is on a fairly basic level, and I appreciate everyone's inputs.

I want to repeat again that I am not on any "side" in the global warming debate. And my opinion is that anyone who claims to be absolutely sure about it may be politically correct but never scientifically correct. There are no scientific proof, one way or the other. Science is not democratic, so counting the forces in your camp is merely silly.

I do stand corrected on the $300W/m^2$ calculation, it turns out to be about -3.3°C. It does not change the point I was making though.

I also understand the objection when I say "reflected" regarding the radiation from that is returned to the Earth from the atmosphere. Anything WILL reflect, unless it is a blackbody, but true, in IR where I am familiar only about 5% is reflected from water or ice. The rest is absorbed and re-radiated. But my expertise, as I pointed out is mainly down on Earth.

Water is an amazing substance and the more I study it the more it amazes me. It has several unique features, among them its incredible ability to store heat, especially latent heat. Just the simple fact that it is a dipole and therefore orders the molecules when it freezes in such a way that the density goes down (unique!) is an important fact that is overlooked. If not for that, we could forget about fish in most freshwater lakes up north, for example.

But the most important thing with water is its ability to moderate temperature here on Earth. The greenhouse effect as such is not a threat—it is an absolutely necessary condition for life on Earth. It creates the moderated thermal equilibrium that allows us to live on a planet that does not change its temperatures too much over the day and year. Whether or not we are tinkering too much with that equilibrium is what this whole debate is about, and no proof is yet presented, one way or the other.

I will look through the comments again one by one when I have a little more time to spend and see if I can make any additional contributions.

Thanks Ken, for pulling these things out of me, and thanks everyone for scrutinizing and commenting on our conversation!

Keith Minto: Key point by Mikael Cronholm about the uncertainties:

> it is always easier to calculate what an object emits than what it absorbs, because emission will be spreading diffusely from an object, so exactly where it ends up is difficult to predict.

Mix in variable convection and possible gas density 'layering' and the difficulty in measurement is compounded.

kuhnkat: _Jim, your comment claims elementary physics around the CO_2 molecular bond and transitions. Would you mind explaining whether the transition energy or the bond energy is more important at atmospheric temperatures and why?

Dave Springer: richard verney says:

> CO_2 delays IR finding its way out to space. However, what I do not understand is how this effectively heats up the Earth.

It doesn't heat the Earth any more than insulation in your attic heats your house in the winter. It slows down how fast heat can escape. Since the heat is arriving at the same but leaving at a slower rate this causes the surface to become warmer which then increases the rate at which heat moves across the boundary. A new higher surface equilibrium temperature is thus established. In reality too many other things change too rapidly for equilibrium to ever be attained but the theoretical equilibrium point will rise and thus the target which the system seeks is that much higher.

chico sajovic: Mikael Cronholm, in your paper you say "Radiation is the strongest heat transfer mode." Please elaborate. From my anecdotal experience conduction and evaporation are stronger modes of heat transfer: sweating is better for cooling down than not, blowing on a cup of hot chocolate is better than letting it cool by radiation.

What do you think of the concept of "back radiation" whereby "downwelling radiation" from colder air warms the warmer surface of the Earth.

Dave Springer: Domenic says:

The polished tube method is not exact. It's an approximation.

It might not be exact but when you have a vacuum in the tube and nitrogen in the tube the thermopile output is unaffected but when you have a greenhouse gas in the tube the thermopile output is reduced.

That proves beyond a shadow of a doubt that there's something very different about greenhouse gases illuminated by infrared.

Feel free to explain the difference via some mechanism other than some gases being transparent to IR and some being opaque. Alternative explanations have two defining characteristics: they are always entertaining and never true.

AJB: Mikael Cronholm says:

Water is an amazing substance…

All your Joule are belong to us[118].

---

[118] http://img600.imageshack.us/img600/235/waterb.png

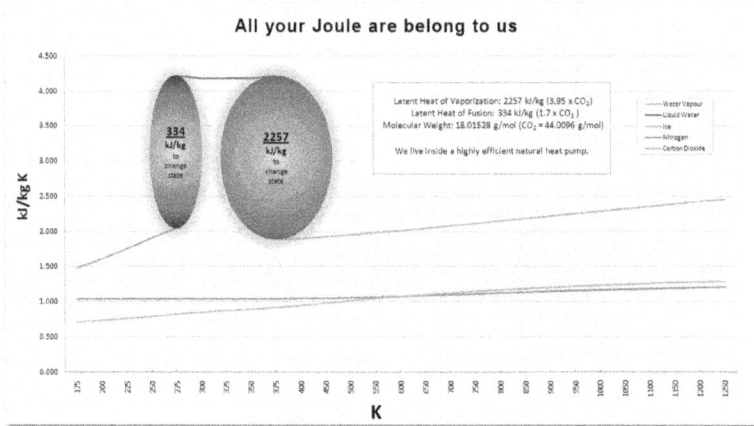

All Your Joule are Belong to us...

Figure 10[119]

Patrick Davis: Awesome article.

> From my way of thinking, the only thing CO_2 can do is increase coupling to space—it certainly can't STORE or TRAP energy or increase the Earth's peak or 24-hour average temperature.

Every AGW supporter I have spoken to cannot get their heads around this concept.

Dave Springer: Mikael Cronholm says:

> There are no scientific proof, one way or the other.

Insofar as some gases absorb infrared radiation and some do not was scientifically proven in the mid-19th century. Your continued denial

---

[119] Ibid.

only means you know less about the physical properties of gases than mid-19th century physicists.

eadler: J. Bob says: W. Falicoff says:

> Further, your statement that it is not possible to achieve a measurement less than the accuracy of an instrument is also not correct, as the accuracy can be increased by taking many measurements (up to a threshold). The resultant accuracy is approximately proportional to the square of the number of readings.
>
> HHmmmm. If a basic sensor has a mean error of 0.75 deg., and 3 sigma of say +/- 0.25 deg., your true error will be between 0.5 & 1.0 deg no matter how many readings you take. You might want to add a few qualifiers.

If the mean error is known, the readings can be adjusted. If the mean error remains constant, than the temperature anomaly, which is what is being sought will not suffer in accuracy. This is the important point, that many who argue about the problems with temperature data seem to ignore, or not understand.

Slacko: Bill Illis says:

> In essence, the Earth has accumulated 1.9 days worth of solar energy. If the Sun did not come up tomorrow, it would take around 86 hours for at least the land temperature to fall below -200C.

What!?? Where I live it can sometimes fall from +40C to -36C between sunset and sunrise.[120] I think maybe you left the atmosphere out of that equation.

---

[120] http://www.wrh.noaa.gov/tfx/pdfs/nws_history.pdf

Tsk Tsk: Dave Springer says:

> *Actually that's not quite right. At low CO_2 concentrations it's nearly linear and as concentration rises it becomes logarithmic. John Tyndall experimentally discovered that too by experiment over 150 years ago. It's pretty much a case of the proverbial low-hanging fruit. The first few molecules get all the fruit they can handle but as the number of molecules increase there's less and less fruit available per molecule. IIRC correctly the curve is logarithmic by the time concentration reaches 100PPM.*

Not sure I like that. The form of the equation shouldn't change for different numbers unless there's a discontinuity. It has to be logarithmic over the entire range of positive real numbers and obviously isn't physical for negative numbers. I think you mean that for low concentrations of CO_2 the dependence diverges only slightly from a simple linear dependence and the divergence grows as the concentration grows.

davidmhoffer: MC—excellent discussion, simple straight forward answers referencing the actual physics. We need more of this in the discussion, not more slogans, so thanks for stepping up!

Now—H_2O vs. CO_2

For all the skeptics who have been trying to present the argument that water vapour is so much stronger an absorber than CO_2 that CO_2 doesn't matter, sorry, but you are wrong. And I am a hard core skeptic!

Water vapour is in fact a much stronger absorber than CO_2 and in the same approximate spectrum. If water vapour and CO_2 were mixed evenly in the atmosphere, that would make CO_2 insignificant. But they are NOT. CO_2 is reasonably well mixed throughout the troposphere, but water vapour concentration falls rapidly with temperature. As a consequence water vapour declines with both altitude and latitude. Whatever CO_2 does or does not do, well over 90% of it exists in the atmosphere at temperatures low enough that water vapour is also low enough that CO_2 becomes significant by comparison.

That said, I repeatedly ask the logical next question that I have yet to see a reasonable answer for from the warmists. If CO_2 reflects/re-radiates/back-scatters whatever you want to call it, upwardly bound long wave, it makes sense that this would result in a temperature increase which in turn would increase water vapour. BUT, and I repeat BUT, does it not also follow that any increase in water vapour works both ways? That is, certainly the water vapour would re-radiate upward bound LW from the Earth's surface, but it would ALSO re-radiate DOWNWARD bound LW too. The downward bound LW that would normally have been absorbed at Earth surface now has an increased % chance of being re-radiate back up. And we're not talking about just the downward LW from increased CO_2, we're talking ALL the downward LW from ALL sources. In other words, increased water vapour may in fact have a net positive feedback, but it has a huge, built in, negative feedback too that likely renders the whole calculation near meaningless.

THERMAL QUENCHING—I saw some comments on this too. This was a big issue for Ernst Beck (may he rest in peace) who felt that this was an underestimated effect of CO_2 LW absorption. His explanation was way, Way, WAY over my head. It was a seriously complex issue as there are so many factors that govern how an individual molecule can or cannot absorb or lose a photon when in collision with another dissimilar molecule, at what temperature, and at what time period before re-emission would happen anyway. I don't have any certainty that he was correct, but the explanations and rebuttals showed one thing pretty clearly—no one has a real good grip on the complexities, in the atmosphere in particular, and measuring thermal quenching is darn near impossible.

Al Tekhasski: For completeness and accuracy of the article it should be noted that the "inexpensive" household IR (non-contact single point) thermometers are designed to measure emission from distant solid objects. Therefore they use the "atmospheric window" where air has the least interference with IR and CO_2 does not absorb nor emit, so they can get better distant readings of objects they are designed to measure. Hence the device uses an IR detector working in a narrow 8-12μm range, right in the center of the window. When pointing to sky, only clouds/haze/smog can affect readings.

Regarding the Antony's picture, he should be able to recalibrate the scale of his FLIR camera to colder side, and it is likely that he would see IR images of clouds behind his house in minus 40 to minus 50 range, where they fly.

wayne: Richard Sharpe says:

> *What you have said, Wayne, does not seem correct to me.*
> *It seems to me that approximately half the energy absorbed by CO_2 would be re-radiated upwards and half downwards.*

Two units of energy goes up, cooling the surface, surface -2, and is absorbed warming the atmosphere, atmosphere +2, your radiators means one unit always goes upward to space, atmosphere now +1, space +1, and one back down to the surface, surface -1, atmosphere is now zero. Please tell me what has just happened, for the surface is -1 unit of energy and space is now +1. Why is this so confusing to you?

And I do know this is ignoring energy that goes via the radiative window directly to space but that effect is exactly the same.

Ian L. McQueen: The following is pretty elementary compared with what has been presented in the text and in the comments, but...this example may help visualization of "back radiation". I shower in the same stall daily. In the summer, I have to run the water a little cool to keep from overheating myself when the temperature is, say, 20°C. In the winter I have to run the water considerably warmer and keep the air temperature around 22°C to keep comfortable. I figure that the difference is due to the difference in temperature of the walls and ceiling in the two seasons.

Related to this, friends heat their house by means of electrical heating panels that warm the drywall panels in the ceiling. Even though they are barely warm to the touch, they are adequate to keep people in the room comfortable. Radiant energy.

Dave Springer: chico sajovic says:

> *Mikael Cronholm, in your paper you say "Radiation is the strongest heat transfer mode."*

Ultimately it's the ONLY mode as convection, conduction, and mechanical transport (evaporation/condensation) ends where the atmosphere ends.

But that is certainly a valid point as we live and breathe and raise our crops on or very near the surface and these mechanisms can accelerate the transfer of heat from surface to above the cloud layer resulting in practically no change at the surface. The increase in temperature caused by increased CO_2 doesn't have to be at ground level. It could be at 10,000 ASL and not affect our surface activities one tiny bit.

In fact, the climate boffins expected to find the CO_2 "signature" as a hotspot in the upper troposphere where the air is very dry and they were confounded when the temperature rise was found to be greatest at the surface where the air is very wet. This should have been their first clue that their climate models were fundamentally wrong. But they'd already decided by then that fossil fuel consumption was going to cause great harm to the planet so they had to continue blaming it.

The mantra morphed from "global warming" first to "climate

change" then when that didn't resonate with the unwashed masses they changed it again to "global climate disruption". That's not science it's a marketing campaign for an ecoloon religion. A floundering campaign, by the way, which is losing ground at an accelerating rate with every passing day.

Tsk Tsk: From the article:

> *KLC: I still feel like I'm missing something. IR heat lamps are pretty efficient, maybe 90%? Let's pick a distance of 1 meter and I want to create a one-square meter flooded with an additional $300W/m^2$. It must be additional irradiation, doesn't it? That's going to take a good bunch of lamps and I would feel this heat. However, I go outside and hold out my hand. It's cold. There's no equivalent of $300W/m^2$ heater in addition to whatever has heated the ambient air.*

What's missing is that the lamps are already radiating based on their room temperature. The $300W/m^2$ from the lamps when they are turned on is, of course, additional, whereas the $300W/m^2$ from the sky is the total heat flux from the sky. The two are not comparing the same thing at all.

It was also nice to see all the talk about the importance of emissivity. It's a subtle point that many get wrong.

Phil: So far the only things that have been discussed are the thermodynamics of CO_2 vs H_2O vapor. What about ICE clouds (slide 23)?

Mikael Cronholm: @chico sajovic, I think you talk about the furnace paper then, and in a furnace radiation is much stronger than any other mode. In the radiation section at the bottom, where the flames are, radiation is completely dominant. Up in the convection section, the tubes draw out the remaining energy they can from the exhaust gases, but that is much less than what is added to the feedstock in the radiation section. The clue is that radiation increases with the temperature to the power of 4, according to Stefan-Boltzmann's law, while conduction and convection don't. Conduction is linear, convection probably less strong than linear, with temperature (logic: just because a surface is 1000°C it does not create a storm around it by convection).

W. Falicoff: My comment above has a mistake. I meant to say "square root" not "square." Let me provide an example. If we take a photo of an astronomical object such as a star or nebula using a CCD camera that has a Signal to Noise Ratio (SNR) for one image of say x, then if we take N images of that same object, the SNR will increase by the square of N. That is the new SNR will be (square root of N) times x.

Mikael Cronholm: @ Dave Springer. I was not discussing properties of gases. I just say there is no scientific proof that increased CO_2 emission causes climate change, or that it does not. And I am not on any side in the debate, for that very reason.

Stephen Rasey: The solar day on Venus is about 582 days.

So it has long nights of 291 days.

It has a surface temp of 735 K, 460° C.

Doesn't that mean it should be emitting 36 times more W/m^2 than the Earth at 300K? $(735^4)/(300^4)$

How is it, then, that Venus remains so hot, even on its night side?

It is closer to the Sun, but shouldn't that mean it is only getting a little over 2.2 times the energy per unit area? Venus also has a high albedo—that ought to help keep it cooler.

Venus has almost 300,000 times more partial pressure of CO_2 than does Earth. It has almost no water. If the IR saturated spectra argument is valid, would 10% CO_2 be just a bad as 99.5% CO_2?

Surface pressure 93 bar (9.3 MPa) Composition:[121]
~96.5% Carbon dioxide
~3.5% Nitrogen
0.015% Sulfur dioxide
0.007% Argon
0.002% Water vapor
0.001 7% Carbon monoxide
0.001 2% Helium
0.000 7% Neon
trace Carbonyl sulfide
trace Hydrogen chloride
trace Hydrogen fluoride

Is it something as simple as PV=nRT?

With the high altitude, low pressure, low temperature stratosphere being the governor for heat loss rate?

W. Falicoff: I realize that many of you are not fond of the website realclimate.org. However, there are useful posts on the above subject (CO_2 radiation exchange in upper atmosphere)[122] and more importantly (saturated gasses).[123]

Also an important data source for the absorptivity of CO_2 is given in the data found in HITRAB.[124] This software uses data from HITEMP.[125]

The HITRAN database has been recognized for more than 20 years as the international standard compilation of spectroscopic absorption parameters for atmospheric gases.

---

[121] http://en.wikipedia.org/wiki/Venus
[122] http://www.realclimate.org/index.php/archives/2007/06/a-saturated-gassy-argument/
[123] http://www.realclimate.org/index.php/archives/2007/06/a-saturated-gassy-argument-part-ii/
[124] http://www.cfa.harvard.edu/hitran//
[125] https://kb.osu.edu/dspace/handle/1811/13476

This paper has some of the CO_2 data available on line).[126]

From what I have read the emissivity of CO_2 does vary as a function of both pressure and wavelength, but I am not an expert on this subject.

DeNihilist: Mikael, I am sitting in front of my wood burning heat-o-lator. When the door is closed and the fan running, the room heats up quite quickly, but my body doesn't. If I want a quick kick of warmth, I turn off the fan, open the door, re-arrange the wood so that the ember side is now facing out, and Whammo! Instant warming of the body.

Also reference the campfire effect. Side facing campfire feels warmth, as the body is cooler then the flames/embers. side away from campfire feels cool/cold, as body on that side is radiating heat to a lower temp atmosphere.

Slacko: Jim D says:

> The measured IR from the sky is pretty much all emitted by CO_2 and H_2O molecules...

No Jim. If that were true, then the warmth of the Sun on my face would be coming in from all directions equally. Or, do you mean at night?

And do you really mean to tell us that nitrogen can't radiate IR?

Tsk Tsk: Mikael Cronholm says:

> @chico sajovic, 8.20. I think you talk about the furnace paper then, and in a furnace radiation is much stronger than

---

[126]

http://faculty.uml.edu/robert_gamache/papers/Rothman_et_al_Preprint.pdf

any other mode. In the radiation section at the bottom, where the flames are, radiation is completely dominant. Up in the convection section, the tubes draw out the remaining energy they can from the exhaust gases, but that is much less than what is added to the feedstock in the radiation section. The clue is that radiation increases with the temperature to the power of 4, according to Stefan-Boltzmann's law, while conduction and convection don't. Conduction is linear, convection probably less strong than linear, with temperature (logic: just because a surface is 1000°C it does not create a storm around it by convection).

Be careful. Just because radiation goes as the 4th power of temperature doesn't mean it's always the dominant form of heat transfer. It also has an extremely low coefficient compared to that of conduction or convection of most materials. Also, both conduction and convection are linear with temperature but their respective ratios are dependent on the particular fluid's properties.

Finally, that 1000°C surface may not create a storm, but it most certainly will drive noticeable amounts of air or any other surrounding fluid flow. Of course, if you're using your hand to measure that then you're probably more worried about the burns you're getting by holding your hand too close to a very hot object.

Mikael Cronholm: Ian W says:

When water vapor in the atmosphere condenses into liquid water and then changes state again and becomes ice, it gives off latent heat for both state changes.

Does that latent heat release follow Stefan-Boltzmann's radiative equation?

No Ian, latent heat is not really related to S-B. Latent heat means that when you add or remove heat, energy, from a substance it will usually change its temperature, except when there is a phase change. At a phase change, for example melting or evaporation, you will be able to add or remove heat without a change in temperature taking place; the change of heat in the substance causes the phase change to take place instead.

In the case of freezing water for example, there is molecular kinetic energy stored in translational movement, i.e. the molecules move around relative to each other (they have a mass, they move=energy). When the liquid turns to a solid crystalline structure, that energy of movement must be removed before the molecules can stand still in relation to each other. That is the latent heat that is released from the water when it freezes. Fruit orchards are actually sprinkled with water to prevent the plants from getting damaged on cold clear night in the spring. The water freezes and gives off heat to the plant, preventing it from getting too cold.

So, as far as S-B is concerned, this process will only influence the input T in that equation, insofar as the cooling or heating of the substance reaches a plateau when latent heat takes effect. The radiation will still always depend on the temperature and emissivity of the substance.

Domenic: To Dave Springer: In actuality, to my knowledge, the true greenhouse effect of any gases (including Nitrogen) has never been measured properly. They are calculated based on a lot of basically untested assumptions. (To explain all the assumptions currently used would require a very detailed and long technical paper.)

To do it correctly, you should use a radiative source at room temp to represent the typical Earth radiation temperatures instead of a high heat source used historically in gas detectors. Then the detector itself should be near absolute zero to simulate outer space at night time rather than the room temperature state it normally is in a gas detector.

In other words, this proposed setup should be an exact simulation of night time conditions. The gases you put inside the tube can represent any kind of atmosphere conditions, or gases, that you wish.

THEN you DIRECTLY measure the longwave radiation transmission, to a great extent, etc of the various gases in the atmosphere.

For those of you with IR guns or devices, you can see true greenhouse effects directly at night by aiming your IR device at the center of the sky on a cloudless night, and then comparing that reading to another night by aiming at the center of the sky when the sky is full of clouds. That is a TRUE greenhouse effect differential measurement.

Slacko: hotrod (Larry L) says:

> *It is perhaps a subtle distinction for some folks, but the CO_2 and water vapor "back radiation" does not "warm the ground"*

Why not? Does it have to do with the 10 micron wavelength or the ocean's surface?

Why not? Those photons have to go somewhere. Why does this look like a "missing heat" problem?

It only "reduces the rate at which it is cooling by IR radiation" to the apparent surface of the very, very cold sky.

So "back radiation—only reduces—cooling."

Huh??

Mikael Cronholm: @ Tsk Tsk. No, but at high temperature it is definitely dominant, and my paper deals with furnaces where temperatures are very high.

Another point to make about that is that radiation and conduction are the most predictable and easy to calculate of the three, at least if conduction is in a solid. Convection is a nightmare!!!

Oliver Ramsay: chico sajovic says:

"Back Radiation" is such a horrible analogy, description or way to think about heat transfer in the atmosphere, that I wish it would just go away. What we are concerned with is the flow of heat from the surface of the Earth to space. It's not that the surface warms the CO_2 in the air then the CO_2 "back radiates" re-warming the surface, it's that when the CO_2 in the air is warmed the flow of heat from the surface to the warmed CO_2 is reduced, thus increasing the temperature of the surface until the flow of heat returns to equilibrium.

I don't quite get the preoccupation with "the surface" when it's actually the air temperature that we measure, not the ground. GHG's don't make the ground warmer in the daytime than insolation is able to do and at night, the paltry amount of energy accumulated in the surface doesn't provide much residual warmth to the air, especially since it's only a fraction of outgoing LWR that is absorbed.

The heat capacity of the air is not enormous but radiative heat loss is not really rapid at these temperatures, either. Something like that?

P Wilson:
In answer to the question up there, the S-B equation doesn't work with dimensional gases.

Secondly, it is a theoretical equation that doesn't work with climate generally.

Practical demo: the basal human metabolic rate is around $58W/m^2$. An average human is around $2m^2$, so the average energy a human generates is around $110W/m^2$. This creates more heat than does the atmosphere or the ground at night under the view of spectroscopes. It is therefore safe to assume that there is much less than even $50W/m^2$ re-radiation.

True, there may be air currents rising, but these are invisible to CO_2 which absorbs (well, it doesn't really absorb but delays by a billionth of a second) energy at around 15 microns, which corresponds to subzero temperatures.

It then quickly thermalizes with nitrogen and oxygen at its most active region—which is quite high in the atmosphere where freezing temperatures correspond to CO_2 absorption. However, at this height (around -28°C in the troposphere) it also competes with the peaks of nitrogen and oxygen absorption (they absorb radiation too).

It is questionable that CO_2 is even a greenhouse gas in real observable terms.

Mikael Cronholm: Slacko, perhaps I can explain. If you look at the beginning of Ken's and my conversation I mention something we call "apparent temperature". It is the blackbody equivalent temperature that something radiates. We use it to determine the reflected radiation in commercial IR measurements. That is also what would be measured in that IR image at the top if the camera was able to measure that low (and emissivity was set to 1 and distance to zero).

So, consider the exchange of heat between Earth and the sky. It depends on the balance between incoming and outgoing radiation, the net difference is the gain or loss. If you have a very clear sky, the sky will have a low apparent temperature, so the temperature difference between the Earth and the sky is the greatest. (You will still have a big influence from the atmosphere; otherwise the apparent temperature of the sky would approach absolute zero. -273°C.)

Clouds will have a higher apparent temperature, meaning that they will radiate more towards the Earth than a clear sky would do, so the difference is smaller and hence the heat loss is also smaller, than with a clear sky.

So, yes, it is true that radiation from the clouds will prevent cooling of the Earth, allowing the Earth to keep its heat to a larger degree. But as the clouds will not be hotter than the Earth, they will not reverse the heat flow from the Earth to become a gain rather than a loss, they can just make it less of a loss. The Sun will give the positive contribution.

P Wilson: Heat leaves Earth by convection. Clouds prevent convection. They don't radiate energy.

randomengineer: Falicoff said:

> ...*further, your statement that it is not possible to achieve a measurement less than the accuracy of an instrument is also not correct, as the accuracy can be increased by taking many measurements (up to a threshold). The resultant accuracy is approximately proportional to the square of the number of readings.*

Apples and oranges, and utter nonsense.

What you describe is simple measurement repeatability with an assumed constant source and then averaging out many multiple readings. All modern measurement devices derive their accuracy spec (traceable to NIST etc) in this manner. Of course, I know this because I made measurement equipment for many years—the measurements in question are those of thermometers looking at a different condition each time, not a fixed, constant source, and the result is that the error is the stated resolution of the device.

If a 1880s era thermometer was good to ± 1 degree, that's your error range. You can't average out 220X 1880's era thermometers and derive accuracy tighter than ± 1 degree.

Konrad: @richard verney I believe this is a valid question. Dave Springer's response fails to address the question in ignoring the diurnal cycle of solar energy input for a given location on the Earth's surface. For CO_2 to cause CUMULATIVE warming, the near surface temperature just before dawn for a single point on the Earth's surface would have to be greater for a local air mass with greater CO_2 concentration than for one with less. A simple empirical experiment could clarify this. I would be interested if anyone could point me to the results of such an experiment.

AusieDan: To: Richard Verney. You asked what sort of time delay does CO_2 impose on the re-emission of the Sun's energy back into space.

I do not know the answer, but suggest we consider what happens when the hot summer ends and chilly winter appears? What happens when day follows night? What happens to the measured temperature at individual locations for well over a century, when you are able to adjust for the rising level of UHI? Not much—that's my simple answer.

JER0ME: John S says:

> the relatively small added heat from CO_2 creates a positive
> feedback loop that includes adding more water vapor into the
> atmosphere until there comes a 'tipping point' where the
> global temperatures go into thermal runaway.

How can it be explained that ANY increase in temperature does not cause this "thermal runaway" 'tipping point'? If the temperature as recently as the MWP was higher than today, which it certainly was in at least some areas and possibly all, why was this mythical tipping point not reached? As it was not, what makes you believe it will be during the present warm period? Is there any evidence for this? Models are NOT evidence, BTW, as they ONLY produce what they are programmed to produce.

P Wilson: chico sajovic says:

> When I am hot I should:

> stand outside naked on a low humidity night

or

get soaked in water and stand in front of a fan

from what we're told, you should do c) go to a special room that has 280PPM CO_2, regardless of its temperature.

davidmhoffer: Slacko says: hotrod (Larry L) says:

> *It is perhaps a subtle distinction for some folks, but the CO_2 and water vapor "back radiation" does not "warm the ground."*
> *Why not? Does it have to do with the 10 micron wavelength?*

ANSWER: Because the downward re-reradiated LW is MUCH more likely be absorbed by water vapour or other absorbers and be re-radiated back up than it is to ever hit the ground. Double CO_2 instantly, wait for a new equilibrium to be established, and you have the EXACT same amount of SW going in, and the EXACT same amount of LW coming out. The EFFECTIVE black body temperature of the Earth is what the IPCC claims is going to rise by 1 degree, and this is at a point high up in the atmosphere (about 14,000 feet if I recall) not at the Earth's surface.

QUESTION

or the ocean's surface. Why not? Those photons have to go somewhere. Why does this look like a "missing heat" problem?

ANSWER

LW radiance cannot penetrate more than a micron or so of water before being absorbed. The result being that any longwave that does strike water is absorbed in a layer so thin that it immediately evaporates taking the extra energy from the LW, plus any energy that was already in that water with it into the atmosphere. Any temperature changes in the ocean have to be attributed as a consequence to other factors such as fluctuation of SW from the Sun, rainfall, runoff and so on. As for "missing heat" that's a measurement problem.

Ken Coffman

QUESTION

It only "reduces the rate at which it is cooling by IR radiation" to the apparent surface of the very, very cold sky. So "back radiation—only reduces—cooling." Huh??

ANSWER—that is an imperfect way of thinking about it, but at day's end it is a fair description. A given photon might, in theory, be radiated upward and go straight out to space. Or it might hit one CO_2 molecule, be absorbed, and then re-radiated in a random direction. Up, sideways, down, whatever. For rough figuring, call it 2/3 up or sideways and 1/3 down. So no matter how much CO_2 you have, the end results is always more ups than downs, and the photon eventually escapes to space, but not have hundred, to millions of absorptions and re-radiations. So adding CO_2 does not add a single additional photon, not one, to the equation. All it does is increases the average number of zig zags through the atmosphere before eventually escaping to space. So yes, extra CO_2 would add no extra heat at all, it would just increase the amount of time it takes any given photon to escape, and "slow down the escape" could roughly be equated to "slow down the cooling."

cal: Slacko says: Jim D says:

> The measured IR from the sky is pretty much all emitted by CO_2 and H_2O molecules...
>
> No, Jim. If that were true then the warmth of the Sun on my face would be coming in from all directions equally, or do you mean at night? And do you really mean to tell us that nitrogen can't radiate IR?

Sorry, Slacko, but Jim D is right. The Sun emits energy in the visible and UV part of the spectrum and a good job it does too. The atmosphere is almost transparent at these wavelengths so unless it gets reflected by clouds or the surface it will make it to the ground where it will be absorbed to warm the Earth. The Earth on the other hand emits at longer wavelengths with a peak at 10 micron (infrared starts at about 0.8 micron) and a lot of this does get absorbed by the CO_2 and water vapour in the atmosphere before being finally radiated into space by the same molecules.

And no, Nitrogen does not radiate (or absorb) in the infrared.

On a more general point, the picture at the top of this piece is a bit misleading. When you point an infrared camera at the sky at night you will only capture the infrared radiation from molecules (mainly CO_2 and H_2O) in the atmosphere. There is no other significant source of infrared radiation in space to detect. Since these molecules are radiating in narrow bands and a lot of it in the far infrared that the camera may not detect the total energy received by the camera is quite small. As explained in the responses by MC the camera will assume it is looking at a grey or black body with energy distributed across the whole spectrum according to Planck's law. So it will calculate what temperature a back or grey body would have to be to output that amount of energy. Because the energy in the narrow bands that are detected is then spread out over the whole spectrum that calculated temperature will appear much lower than it really is and the sky will be shown as black.

Robert Clemenzi: I agree with Berényi Péter with respect to:

> *For 300 W/m^2 radiation I get -23.4°C at 300 W/m^2 when I calculate it.*

My temperature conversion calculator correlates 300 W/m^2 with -3.3°C (assuming an emissivity of one), and -23.4°C with either an emissivity of 73.4% or 220 W/m^2. A few lines later your body will radiate approximately 648 W/m^2 implies that your skin temperature is at least 129°F (don't think so).

To Jim D:
Yes, IR is reflected. The 324 W/m^2 back radiation from Kiehl and Trenberth (1997) implies a temperature of 34°F if the emissivity is

one, but 58°F (15°C) with an emissivity of 0.8 (-20% in my calculator). Since I don't think that the atmosphere is able to be the same temperature as the surface (remember, the troposphere cools with increasing height), some of the energy must be reflected and not emitted. Some calculations indicate that even 80% is too high—in the absence of clouds. On the other hand, since about 50% of the planet is covered with clouds, it sort of makes sense to assume that some of the energy is "reflected" by the cloud bottoms.

In fact, on cloudy nights, the hand held radiometers indicate an apparent cloud temperature within 2°F of the surface temperature.

In addition to the expected reflection by the droplets themselves, the heat from the surface evaporates droplets at the cloud bottom, which then causes the vapor rise a few inches, where they recondense. For both evaporation and condensation, the associated spectrum is nearly blackbody, with no spectral lines. Notice that this is similar to reflection since the energy emitted is not affected by the temperature of the cloud, but only by the temperature of the surface. ("Scattering" may be a better term.)

For Jim Masterson:
In Kiehl and Trenberth (1997), the downward flux of 324 W/m^2 is specifically for cloudy days, it is 278 W/m^2 for clear days.

To all:
At a single frequency, the change in absorption is logarithmic. However, when the entire IR band is considered, and over the range from 200PPM to 500PPM, the change in CO_2 absorption is logarithmic with $R^2=0.9988$ and linear with $R^2=0.9932$. This is because as each frequency becomes more saturated, a new frequency starts to absorb.

Mikael Cronholm: P Wilson says:Mikael Cronholm says:

> ...heat leaves Earth by convection. Clouds prevent convection. They don't radiate energy

Reply: Sorry Mr. Wilson, but I do not agree with what you say. (To other readers: I was not the one who made the statement above.)

Convection takes place in substances in the form of fluids and the atmosphere itself is a fluid, so convection can take place in it, including within the clouds themselves (they don't "prevent" convection, they take part in it). It is true that heat transfers from the ground to the atmosphere, but it is still part of the heat that surrounds the planet. So assuming that we count the atmosphere as a part of our planet, as opposed to space around it, no heat will enter space by convection because there is no fluid that can circulate into space. Clouds do radiate, because everything with a temperature does.

Let us make this clear. The only way that Earth, and its atmosphere, can exchange heat with space is by radiation.

Don V: I have been fascinated by this posting—as well as all of the many cogent comments from the insightful experts that have been attempting to answer many of the questions that have been posed in the subsequent comments here.

If I may, I would like to add my two cents worth to clear up some things I have read here, and use this forum as a sounding board/perhaps clarify some misconceptions. Please correct me if I am wrong.

First, regarding reflection, vs. absorption/re-radiation. Although, at a molecular level, it may seem they are the same, there is a difference. Corner cube reflection in a prism of glass or reflection that occurs inside a droplet of water generally

1) absorbs incoming light at the first material boundary, transmits through the second material, then reemits the incoming light back into the first material and maintains all spectral integrity and most of the intensity of the incoming light, (polarization may occur)

Ken Coffman

2) there is very little if any loss of energy EXCEPT at those wavelengths that the second material absorbs light and

3) reflection/refraction occurs where either a phase change, or significant density change is encountered. It isn't widely understood but, gases can be seen to refract and change the apparent direction of incoming light and this could be considered "reflected" light, but it is nowhere near as dramatic an effect as what occurs in the reflection of nearly all visible light when it encounters many, many tiny droplets of liquid water (clouds or rainbows). Two examples of air refracting/reflecting light are the Schlieren waves you see when you look at a distant scene across a hot parking lot or runway[127] and the "mirage" you see of a distant city or the reflection of the sky to give the appearance of an oasis of water on a hot desert.

But that is NOT what happens when a molecule like water or CO_2 ABSORBS light at their respective infrared bands. The definition of absorbance is $A = 1 - R - T$ where R is the reflectance and T is the transmittance. Absorbance is the loss of light at specific wavelengths of light, and these unique wavelengths for both CO_2 and water correspond to electronic, vibrational, and rotational energy modes that are unique to each molecule. They happen at unique wavelengths because only quanta of energy that match the exact electronic, vibrational and rotational energies for each of these molecules can cause them to absorb and become excited at these quanta of energy. Generally, if a molecule is excited at a specific wavelength, but it immediately gives up that energy and decays back to an unexcited state and thus reemits that photon, it is no different than transmitted light.

The resulting photon seems to pass right through the

---

[127] http://hiviz.org/hsi/ss/schlieren/index.htm

excited→unexcited molecule. Outside of absorbance bands this is usually what happens to both visible and infrared light when it interacts with gases like CO_2 and water. However, at very short UV wavelengths light absorbance/scatter has the effect of making the absorbing gases appear to "glow" in the visible spectrum—Rayleigh scattering and Mie scattering—which is why the sky appears sky blue, but at sunrise or sunset appears to have a rainbow of colors, and why the Sun appears yellow.

At absorbance bands, however, scatter, transmittance and reflectance don't often happen. The incoming photons with quanta of energy that match absorbance bands of any given gas molecule are absorbed by that gas molecule and it is excited up into a higher electronic or vibrational, or rotational energy state, and if or when it reemits that energy it necessarily must experience some LOSS—entropy gets it's cut. Because the molecule has an "affinity" for quanta of energy that correspond to its natural "frequencies", it is also very rare that light that is "absorbed" at a given wavelength is reemitted at that wavelength. If it were you wouldn't see the loss of light—absorbance—at that wavelength.

Instead, energy that is absorbed at one quanta is usually released only when:

1) even more energy is pumped into the molecule at the same quanta (rare), or

2) it reemits that light at a longer wavelength (fluorescence—does CO_2 fluoresce in the infrared? I'm not sure but I doubt it.), or

3) the molecule collides with another molecule that has less energy (the usual case). When collisions occur, (very frequently) the absorbed quanta of energy is transferred between the two molecules so that they balance out—generally both of them just pick up more translational energy—they move a little faster.

In general devices that "measure" the temperature of a gas or a liquid actually transfer (come into equilibrium with) the translational energy of that gas or liquid to a visual media—indicator liquid that expands in a fixed volume—or transfer the translational energy to an equivalent vibrational energy in a solid material which is used to derive an electronic (thermistor, thermocouple etc.) or visual signal (bimetal). So an observed "temperature" of a gas is an average of all of the molecule's translational energies in a given volume that interact with the surface of the measuring device.

Now having said that, I would like to make the following assertions:

1) As stated by several previous posters, water has many more, and much broader absorbance bands in the infrared than CO_2,

2) Water vapor, and water droplets are much, much more abundant in the lower atmosphere than CO_2,

3) Even at the now greater concentration of CO_2 in the atmosphere the relatively small amount of IR energy it might absorb is quite quickly transferred during collisions to the much greater abundance of water liquid, water vapor, N_2, or O_2 molecules, water vapor, water liquid, and ice all have greater heat capacity than any of the other gas molecules in air. Liquid water's heat capacity is significantly greater!

4) Because water experiences phase changes and significant density changes, it is redistributed throughout the atmosphere in ways that have a significantly greater impact on the transport of energy from the lower atmosphere and planet's surface to outer space than any of the other gases that make up our air,

5) Since water vapor regularly experiences both gas to liquid, liquid to solid, and even gas to solid transitions in the atmosphere:

a) At any given instance the loss of incoming solar radiation energy (on the day side of the planet) by direct reflection, refraction and back scatter, by clouds, snow, rain and the oceans are significantly greater in magnitude than all of the IR absorbed by CO_2 combined.

b) At any given instance the total amount of incoming solar radiation energy that results in phase change of water liquid to water vapor, greatly and significantly exceeds all of the IR energy absorbed by CO_2 combined.

c) At any given instance the total amount of energy contained in water vapor, water liquid droplets (clouds) and water ice particles (high clouds) in the planet's atmosphere by a huge amount dwarfs the total amount of energy contained in IR excited CO_2 gas molecules, so much so that I would think (I have no proof) that most of CO_2's translational energy content is created more by collisions with water molecules than IR photons by a very large ratio.

d) And finally and most importantly, changes in the concentration of CO_2 that have been observed have gradually been increasing over DECADES. But in any given DAY the concentration of water vapor, water liquid and ice particles in the atmosphere change by orders of magnitude greater concentration than the few hundred parts per million that CO_2 has changed. These significant concentration changes even at the poles have a far greater effect on both immediate weather and of course long term climate. Alarmists would have us believe that water cycle does not now have the capacity to mitigate and regulate the small change in CO_2 concentration that have occurred after centuries and centuries of significantly higher CO_2 concentrations and significantly LOWER CO_2 concentrations in the past.

Feet2theFire: @chemman:

> ...since molecules are 3-dimensional and constantly in motion "green house gases" will re-radiate IR energy in 3 dimensions not just back at the Earth like a reflector.

I've always assumed they all deal with it as 3D re-radiation—including only the downward re-radiation in the heat flow system. Otherwise some of the skeptics would have certainly ripped into them about it.

Steeptown: This is a very useful discussion, particularly for those not of a physics bent. As someone who has worked for many years in the field of fluid flow and heat and mass transfer, it is evident to me that the radiative effects of CO_2 in the atmosphere are of 2nd or 3rd order compared to the radiative, convective and latent heat effects of H_2O. Our climate is dominated by the water cycle in all its forms.

Feet2theFire: I have some hands on industrial R&D experience with heat flow vs. temperature, and also with the reality of emissivity.

I ran experiments using what are called "engineering plastics" in order to find the best plastic for a 420°F application. In one, we heated the plastic to temperature and then subjected it to weight loading to see what kind of strain we got. In the process, I handled up to 500°F pieces of plastic (a bit higher than the real application, for "cushion"). I did this barehanded. Even with a thermocouple telling me the plastic was 500°F, the plastic was almost not even warm in my hand. It is not the temperature that burns; it is the heat flow. See the next paragraph...

As to emissivity, we also had highly polished and chrome plated tool steel (H13) that was oil-heated to the 420°F temperature. The polishing and chrome plating made the emissivity extremely low (how I still don't "get"). I could put my skin well within 1/16" of the surface of the hot metal and not even feel warmth. (It is not easy to do without jittering, but I did manage to do it, bracing my hand on something cooler, just to see.) However, once the heat transfer method went from radiation to conduction—when I actually touched the metal—it was instant BURN!

Low emissivity is amazing.

Low heat flow is also amazing.

Note: By far the best fix for the occasional burns was using Freon from a spray can. I was SERIOUSLY [snip . . irritated?] when they banned CFCs to protect the ozone hole. And I completely believe that in time that theory will be completely refuted.

Man BearPigg: Domenic: the CO_2 measurements by Scripps Institute are taken in several locations around the world including the locations near the North and South Pole.[128]

> These measurements support the findings of the measurements at Mauna Loa. They also show there is a lag of CO_2 from the Northern to Southern Hemisphere (the levels in Antarctica lag those in the Northern hemisphere), as expected given the gains in CO_2 are primarily from anthropogenic sources.
> —W. Falicoff

> For a theory to be scientifically proven, it has to be stipulated and tested, and the test must be repeatable and give the same results in successive tests for the theory to be proven.
> —Mikael Cronholm

---

[128] http://scrippsco2.ucsd.edu/research/atmospheric_co2.html

So all you have to do now, W.F., is show your repeatable science and you will have done something that would change the entire scientology of AGW.

P Wilson: Mikael Cronholm says:

> Re convection: it is known as convective inhibition when cloud cover at night=warmer than cloudless night.

Clouds form a barrier. From IR it blocks some solar radiation, and from Earth it prevents convection currents of rising air from leaving. Most energy leaving is via convection. It is known as atmospheric convection.

ThomasU: Very interesting post & posts! Great source of knowledge down to little known details (such as spraying water on fruit). Lot of food for thought!

I only have one question to add to this conversation: Is it at all possible to measure the energy balance (or radiation balance) of the Earth. By energy (radiation) balance I mean the difference between energy input and output. Please excuse if this question is put forward in terms which fall short of physicists standards—that's because I am just a curious layman.

P Wilson: To clear the matter: Since IR radiation from Earth is weak, it can't move through the atmosphere and is absorbed by water vapour mainly, and CO_2.

Since warmer air is less dense than cold air convection occurs to take energy upwards, via what they call advection.

wayne Job: We have a little blue planet that is stirred internally by a big yellow monster, this same monster bombards us with the full spectrum of radiation. Mitigating factors protect us from this monster or we would perish. Likewise if it failed to arise tomorrow we would perish very quickly. Our little blue planet works its fanny off trying to balance an out of balance heat input caused by all sorts of wobbles and perturbations. The occasional large volcanic eruption is compensated for and the world returns to normal. A flea on elephant comes to mind as to effect of CO_2 on the planetary climate. The cyclic changes we see if we live long enough are the result of the rules of thermodynamics in an open heat pump trying its utmost to balance itself.

It would be pure folly and ego on our part to contemplate that we could change the climate. Looking backward to Newton and celestial mechanics using real geometry is the key to finding the reasons behind ice ages and interglacials. When these are understood then we can look at the minor perturbations with some understanding and a modicum of confidence. Some scientists are looking at this aspect and finding a surprising correlation to the Sun's recalcitrant attitude.

stephen richards: I agree with everything MC said except (as Jim D noted) that IR is not reflected by CO_2, it is absorbed and then re-emitted in all directions.

If the CO_2 molecules have radiatively absorbed and are therefore above ground state, they will likely 'reflect' further incoming IR.

P Wilson: ThomasU says:

> *This is indeed a big problem. Temperature measurements are used, which are inadequate, sparse and unreliable.*

Atmospheric energy should be measured in joules per cubic metre. Not in temperature...

paulhan: I'm not understanding this idea that a 1deg rise in temps caused by CO_2 will cause a further 2deg rise through the extra water vapour that is created.

Surely if that were true it would be possible to surround a

small lake with something like the Eden Project biodomes, pump it full of CO_2, and then extract the heat that all the extra water vapour created. It would be the ultimate passive energy creator, no windmills, solar or thorium required. At the very least we could do it on a small scale to test the thesis that that is what would actually happen.

Bomber_the_Cat says: Steve Reynolds says:

> *I agree with everything MC said except—that IR is not reflected by CO_2, it is absorbed and then re-emitted in all directions*

To which Latitude asks:

> *Steve, Since it's constantly being exposed to IR, when it's saturated, does it just reflect because it can't absorb any more? Or is it in a constant state of absorbing and re-emitting? neutralizing itself?*

The gases in our atmosphere (N_2, O_2 and CO_2) do not 'reflect' light. If they did it would be impossible for us to see coherent images because the light would be bouncing all over the place. We can only form clear images where light travels in straight lines. The same is true in the infrared. You no doubt have seen pictures from infrared imaging cameras, e.g. police helicopters. These are only viable because atmospheric gases do not 'reflect' infrared radiation but allow it to travel in straight lines.

However, CO_2 does absorb infrared within a narrow wavelength band. The important absorption occurs at wavelengths between 14 and 16 microns (other CO_2 absorption bands are not significant in the context of greenhouse warming because they do not obstruct the radiation emitted from the Earth's surface). Photons of infrared radiation around 15 micron have just exactly the right amount of energy to raise a CO_2 molecule from one vibrational state to another one. Being thus 'tuned', they may be captured by a CO_2 molecule which then changes state. Almost instantaneously, within nanoseconds, the CO_2 molecule re-emits the photon and reverts to its original state. Because the photon can be re-emitted in any direction (spherically random) about half of them will be re-directed back to Earth and will represent an additional heat flux if they ever get there.

If the 'excited' CO_2 molecule collides with another molecule, say of Nitrogen, before it can re-emit its captured photon, then it becomes 'thermalised' (or as Mark Wagner says, 'quenched'), its extra energy is transformed into heat and it can no longer emit a photon. If this happens the atmosphere is warmed.

P Wilson: It's physically impossible for a 1°C rise in temperature by CO_2.

What causes temperature changes are weather systems (via convection). Storms, snow, cyclones, clouds—the whole gamut are caused by convection. When air convects, it cools sufficiently for it to fall below its dew point. That's when clouds form.

I don't know why CO_2 was brought into meteorology and climatology.

Maybe it was to paint a human face on climate.

Bomber_the_Cat says:

> *What you haven't mentioned is the temperatures at which this 14.77 peak of CO_2 absorbs radiation. It corresponds to -28°C which is up in the lower troposphere. When it leaves that temperature range it thermalises.*

There is no physical mechanism that can cause subzero temperature ranges to penetrate back to Earth and warm it up beyond the nominal 14-15°C. That is tantamount to saying that putting cold water in the freezer will cause it to be lukewarm.

Smoking Frog: Jim Masterson The cloud cover is supposedly 62%. KT 1997 combines three cloud layers (49%, 6%, & 20%) to get that figure. They call it "random overlap," whatever that means. It looks like an application of the Inclusion-Exclusion principle. I get 61.6% which rounds to 62%. Their famous energy diagram (Fig. 7 in KT 1997 and Fig. 1 in TFK 2009) should state: "62% cloud cover assumed."

The Inclusion-Exclusion Principle requires that we know the overlap(s) to begin with, so it can't give us the answer. The average of the minimum and maximum overlaps gives exactly 62%:

$$((49 + 6 + 20) + 49) / 2 = 62$$

but I'm not sure that this is what we'd get with random overlap, even if, as my calculation assumes, there are no real-world constraints.

AJB: Don V says:

> An excellent post, Sir. IMHO non radiative transfer of energy to the tropopause by water completely swamps any radiative effects of increasing CO_2.

What happens at and above this point is more interesting, however.

Bomber_the_Cat: P. Wilson says:

> What you haven't mentioned is the temperatures at which this 14.77 peak of CO_2 absorbs radiation.

The CO_2 absorption of photons at 15 micron is not dependent on temperature. A molecule of CO_2 will absorb any photon with the 'correct' quanta of energy which impacts it, irrespective of temperature.

More interestingly:

> *There is no physical mechanism that can cause subzero*
> *temperature ranges to penetrate back to Earth and warm it up*
> *beyond the nominal 14-15°C.*

I assume that subzero here means below zero on the Celsius or Fahrenheit scales. Now ALL objects above absolute zero emit radiation; that includes you, me and an ice cube. The amount an object emits (Stefan's Law) and the wavelengths of the radiation emitted (Planck's Law) is determined by their temperature. So even cold objects, such as an ice cube, emit radiation. If that radiation impacts a warmer object then it is absorbed by the warmer object. The warmer object therefore receives energy that it wouldn't receive if the cold object was not there—and so the warm object is kept warmer than it otherwise would be. Radiation is not somehow preferentially attracted only to colder objects. Thus the presence of cold objects can keep warm objects warmer!!! This does not infringe the 2nd law of thermodynamics, only a schoolboy misunderstanding of it.

As for making the Earth warmer, remember that the Earth is warmed by the Sun. The CO_2 'blanket', by sending radiation back to the surface, simply acts as insulation. The incoming solar energy is what causes the Earth's surface to rise.

At the top of the atmosphere of course, the amount of radiation leaving the Earth will always balance the radiation coming in—no matter how much CO_2 there is in the atmosphere. But we don't live at the top of the atmosphere, we live on the surface—and it is the surface temperature which is affected by the greenhouse gases.

David L: W. Falicoff says:

> *Further, your statement that it is not possible to achieve a*
> *measurement less than the accuracy of an instrument is also*
> *not correct, as the accuracy can be increased by taking many*

measurements (up to a threshold). The resultant accuracy is
approximately proportional to the square of the number of
readings. This holds true in several fields.

Actually, repeated measurements cannot improve accuracy, it can only improve precision. If there is bias in the gauge then repeated measures will keep giving you the biased answer. Precision (e.g. the standard deviation) of the average value of the measurement can improve by the inverse square root of the sample size.

But there's far more to it than increasing sample size. Check out MSA (Method System Analysis) or Gauge R&R (Repeatability and Reproducibility). In a manufacturing environment a Gauge (or measurement system) is not considered capable until the precision is at least 1/10th the specification range.

So in terms of AGW, if the world is going to melt down at an additional 2°C then the gauge should be able to repeatably and reproducibly measure to ±0.2°C on the individuals—not multiple measurements to construct an average.

izen: davidmhoffer says:

> *LW radiatiance cannot penetrate more than a micron or so of*
> *water before being absorbed. The result being that any*
> *longwave that does strike water is absorbed in a layer so thin*
> *that it immediately evaporates taking the extra energy from*
> *the LW, plus any energy that was already in that water with*
> *it into the atmosphere.*

The LW absorbed in the surface layer may not be enough to liberate a water molecule from the bonds at the surface layer, it may just increase the random kinetic energy of the molecules in the surface layer that is transported to the deeper layers millimetres further down by surface turbulence. Water is rarely so still that this transport mechanism is insignificant.

Even if it does provide enough energy for a water molecule to break free of surface bonds and 'evaporate' that molecule has a ~50% chance of colliding with the other molecules and bouncing back into the water surface adding to the thermal kinetic energy of the surface layer.

There is no way that the water surface acts like a one-way street for all incident radiant energy. It is not Maxwell's demon!

Dave Springer: P Wilson says:

> what you haven't mentioned is the temperatures at which this 14.77 peak of CO_2 absorbs radiation...

The temperature of the gas has little to do with IR absorption. The Earth emits LWIR in a continuous blackbody spectrum with a peak emission at around 10μm which corresponds to a blackbody at approximately 52°F which is the average surface temperature of the ocean. CO_2 absorbs narrow bands out of that continuous spectrum which excites the CO_2 molecule and because the molecule is part of a cold dense mix of gases the excited molecule almost instantly bumps into a neighbor (most likely N_2) which then thermalizes the N_2.

Re-emission up the upwelling narrow band energy is a continuous blackbody spectrum corresponding to the temperature of the gas at whatever altitude the re-emission occurs. The process starts from the ground up and proceeds to saturation. Looking down at the atmosphere from above with a spectrometer one sees a continuous blackbody spectrum with narrow absorption bands where the energy level falls off. The energy missing from those narrow bands is redistributed across the rest of the spectrum. The altitude/temperature to which you refer is the energy level at the top of the 15μm band.

See Figure 8.2 of *A First Course in Atmospheric Radiation*[129],

---

[129]   http://www.sundogpublishing.com/AtmosRadFigs.html, *A First Course in Atmospheric Radiation*, Grant W. Petty, Sundog Publishing, 2006.

spectrograph from 20km looking down on the North Pole.

The blackbody emission curve is at about 265K which corresponds to the surface temperature at the pole at the time. You'll notice a big hole in the spectrograph centered on 15μm with the bottom of the hole following the 225K blackbody curve. The missing energy at 15μm is what CO_2 has absorbed beginning from the ground up and has been completely thermalized by an altitude where the air temperature is 225K which, applying the dry adiabatic lapse rate of 10K per 1000 meters is about 3000 meters or 10,000 feet (middle region of the troposphere).

The missing energy in the 15μm, when thermalized, is re-emitted as continuous blackbody spectrum so the top of the curve at 265K (which is the surface temperature) is a bit higher than it would be otherwise. That bit higher surface temperature is the effect of CO_2's action as a greenhouse gas. It impedes the flow of 15μm energy from surface to space which in effect acts like insulation making the surface temperature a little higher than it would be otherwise. The temperature of the cosmic void is about 3K and doesn't change. The increased differential between surface and space raises the rate at which energy flows from surface to space re-establishing a new surface temperature equilibrium point between energy-in (short wave energy from the Sun) and energy-out (long wave energy from the Earth).

This is all undisputed except by cranks and other assorted ignoramuses who refuse to accept the radiative absorption and emission characteristics of various gases according to physics theories that have been around for over 200 years and which were experimentally confirmed 150 years ago by John Tyndall and which have remained as well established as any theory in physics since then.

The controversy isn't about the direct effect of increased CO_2. That's cut and dried number crunching of basic physics formulas. The bone of contention is in the feedbacks. The climate boffins on the "hockey team" insist there is a large positive feedback which will somehow cause a runaway greenhouse despite the fact that all paleo-climate evidence of every kind shows the Earth has never in its history experienced a runaway greenhouse despite atmospheric CO_2 levels far higher than could be obtained by burning every last drop, wisp, and crumb of recoverable fossil fuels.

ALL the evidence says the feedback is negative which limits the maximum global surface temperature and where the temperature increase (we're in a very COLD period of the Earth's history) is concentrated in the higher latitudes i.e. the tropics won't get much warmer but the temperate and polar regions will. The usual state of affairs for the Earth over the past billion years is warm and friendly for living things from pole to pole.

The current terrestrial biosphere is a shrunken frozen shadow of itself much of the time over much of its extent compared to the warm (non-ice age) periods which is to say 90% of the time for uninterrupted periods lasting as long as hundreds of millions of years. The Earth has been in an ice-age for the past 3 million years and ice ages are not the normal state of affairs—they are the exception to the rule of a planet lush, warm, and green from pole to pole with far higher atmospheric CO_2 content and a far larger/faster carbon cycle driven by living things.

NoIdea: Hello, Izen. Quote:

> *There is no way that the water surface acts like a one-way street for all incident radiant energy. It is not Maxwell's demon!*

But CO_2 is a demonic gas, with its wave number of 666.666 (15µm)!

Is it not Maxwell's demons casting back all the IR radiation?

The blackbody temperature to emit a peak IR of 15 µm is according to Wien's law 193.18456 Kelvin. This equates to -79.97 Centigrade (please do not miss the MINUS sign, yes nearly minus 80°C!).

How hot can this 15µm -80 Centigrade IR heat anything?

P Wilson: Bomber_the_Cat says:

> *Thanks for the lesson on radiative physics theory—I'm quite*
> *aware if it. (The theory, that is)*

Which is all that it is. A theory.

In theory, I receive energy from my surroundings, but they never surpass my basal metabolic rate of $58w/m^2$. Similarly I transmit to cool night air energy at this basal rate, although only to the immediate few inches at most, and this energy soon thermalizes (disappears as the form of heat. Heat is not a permanent).

There's little point in invoking the rather absurd S-B equation, which is a thought experiment that leads to rather absurd conclusions.

The CO_2 blanket is a nonexistent phenomenon. It's a theoretical phenomenon—since it paints a human face on a climate system, just like Big Bang paints a human face on the unknown origins of time and the universe.

It was established well before AGW ideology that CO_2 delays the transit of 8% of total IR radiation, regardless of its quantity. As you know, as 14.77 microns the saturation window closes. 14.77 microns does broadly correspond to -28°C. 10 microns corresponds to around 15°C, the average global temperature. So the last place that CO_2 absorbs radiation is at the surface, which on average is above subzero. CO_2 is invisible to these temperatures.

In climatology the fix is made by saying that something is going up in the upper levels of the lower troposphere where indeed these subzero (°C) temperatures occur and where CO_2 captures radiation at its peaks. Only it really isn't, and what radiation is absorbed at subzero temperatures is quickly released.

There is no radiative mechanism by which -28°C can increase to 15°C or upwards from a base point of -28°C. Unless, of course you increase the heat source (The Sun). But then CO_2 becomes invisible again to SW radiation (It absorbs/delays at subzero).

If you want to prove that an ice cube will heat a cup of coffee—a tepid one, to give you the benefit of the doubt (which is the gist of your argument—that used by alarmists to prove the CO_2 conjecture) then we'd all be grateful for your video test results.

Dave Springer: Nice. It's heartening to see some commenters who understand the physics. "School-boy" misunderstanding might be a bit harsh but I understand where you're coming from. A good understanding of the four years of science taught in grades 9-12 covers almost everything you need to know about this subject in biology, chemistry, and physics from how the greenhouse gases work to what acid rain from sulfate emissions does to how these thing effect the biosphere. When PhDs get involved in the conversation they start such a pedantic bickering over small details in their fields of expertise that the big picture gets lost and laypersons then proverbially can't see the forest because there are too many trees blocking the view.

izen: P Wilson says:

> ...heat leaves Earth by convection. Clouds prevent convection.
> They don't radiate energy...

Some energy leaves the SURFACE by convection, but it does not get very far. It is not clouds that block it—strange idea! It's the adiabatic lapse rate.

Convection can only move air until the increased buoyancy from the lower number of molecules per cubic metre is offset by the lower density with increasing altitude. It is the lower density of the atmosphere with altitude that blocks convection.

The low altitude of most clouds indicates that the temperature falls below freezing just a few Km above ground level and the atmosphere regains that latent heat of evaporation well below the tropopause. Above the main low cloud layer water vapor is much less important as an atmospheric absorber/emitter.

Ken Coffman

Dave Springer: P Wilson says:

> As you know, as 14.77 microns the saturation window closes.
> 14.77 microns does broadly correspond to -28°C. 10 microns
> corresponds to around 15°C, the average global temperature.
> So the last place that CO_2 absorbs radiation is at the surface,
> which on average is above subzero. CO_2 is invisible to these
> temperatures.

This is laughably wrong! PEAK emission frequency corresponds to
those temperatures. A blackbody emits a CONTINUOUS spectrum
with broad shoulders that fall off slowly on either side of the peak
frequency. There's plenty of 15µm energy coming from a
blackbody with 10µm peak frequency. You obviously have no idea
what a continuous blackbody spectrum looks like. What you're
saying is like saying no colors of visible light come from the Sun
except for yellow because that's the peak emission frequency of a
5200K blackbody source. Ridiculous misunderstanding.

izen: Hello, NoIdea. As usual you ask a deep question.

> Any blackbody emitter above -80 Centigrade will ALSO emit
> more energy in the 15 µm band, as well as more at shorter
> wavelengths.

But CO_2 is NOT a blackbody emitter. Like the other atmospheric
gases, including water vapor it is a very poor emitter over most of
the spectrum. It just has strong bands of emission/absorption like
water vapor at wavelengths related to its molecular vibrational
modes.

Very hot CO_2 will still have a peak emission spectra in the 15
µm band because that is its surface 'colour', it does not have a
blackbody emission/absorption spectra but one vastly biased to the
energy level of the molecular vibrational modes.

In the infrared longwave spectra CO_2 has a surface 'colour'

197

that modifies its emission spectra; think of it as a 'Deep Purple...'
-grin-

Phil.: Tsk Tsk says:

> Not sure I like that. The form of the equation shouldn't
> change for different numbers unless there's a discontinuity. It
> has to be logarithmic over the entire range of positive real
> numbers and obviously isn't physical for negative numbers. I
> think you mean that for low concentrations of CO_2 the
> dependence diverges only slightly from a simple linear
> dependence and the divergence grows as the concentration
> grows.

No need for a discontinuity, if you expand the terms in the equation
for small $[CO_2]$ you get a linear dependence, for medium values you
get $\sqrt{\ln([CO_2])}$ and for large values $\sqrt{[CO_2]}$. For the range of values
in the atmosphere $\ln[CO_2]$ is a good fit.

P Wilson: Izen, thanks for the reply.

Clouds keep the Earth warmer (or a given portion of the
Earth) by preventing heat from escaping. In fact, convectional
currents can rise a long way into the atmosphere before they lose
their buoyancy. Something like a hot air balloon does.

This uplift and cooling of air is what causes most weather.
Cumulonimbus clouds are often 10,000 metres in height from
surface.

Actually, water vapour above this level acts as an absorber of
radiation: the so-called greenhouse effect.

J. Bob: eadler says:

> If the mean error is known, the readings can be adjusted. If
> the mean error remains constant, than the temperature

anomaly, which is what is being sought will not suffer in accuracy.

And those "if's" can be significant. Much of the mean error "adjustment" ability depends on the sensor, electronics, environment, calibration protocols, and how well these protocols are ACTUALLY followed.

My first "outside" job, over 60 years ago, was taking the hi-lo temperatures, for a neighbor, on an old Taylor mechanical thermometer. It had mechanical slides that indicated the hi-lo points, and if I remember, 2 deg. graduations. The neighbor then sent the data, to some government agency. At $0.50 a week that was big money back then. A long way from 3/4 wire platinum RTD's.

Falicoff's comment about taking multiple readings may "enhance" the resolution, or quantization, but not necessarily the accuracy. This is also known as "resampling", as denoted in p.334, *The Handbook of Astronomical Image Processing*, Berry & Burnell.

Dave Springer: P Wilson says:

How hot can this 15μm -80 Centigrade IR heat anything?

Because it sits between a 15°C ocean surface and the -253°C of the cosmic void. It doesn't heat anything. It slows down the rate of cooling by interposing something warmer than the black of space between the surface and the black of space.

This isn't rocket science. It's about as difficult to understand as is understanding why sitting a cup of hot coffee on a block of dry ice versus a block of water ice. In both cases the coffee won't get any hotter but it will cool a lot faster sitting on the dry ice versus sitting on the regular ice.

Or even better, hot coffee in an insulated thermos vs. a non-insulated thermos. In both cases the coffee is going to get cooler but the insulated thermos will slow down the rate of cooling. To get an even closer analogy consider an insulated versus and uninsulated hot water heater where the heating element is turned on once a day for a set period of time. The insulated water heater will have a higher temperature because the amount of energy added to each one is equal but the rate of escape of that added energy is lower for the insulated vessel. The end result is the water in the insulated vessel will have a higher maximum and minimum daily temperatures. In the case of the ocean it's the Sun doing the heating on a daily basis and greenhouse gases are the insulators which slow down how fast the ocean cools when the Sun isn't heating it. As Ernest Rutherford said:

> *If you can't explain a theory in physics such that a bartender can understand it then the theory is probably wrong.*

A bartender can understand the difference between an insulated and uninsulated thermos. Why can't you?

Oliver Ramsay: Bomber_the_Cat says:

> *If that radiation impacts a warmer object then it is absorbed by the warmer object. The warmer object therefore receives energy that it wouldn't receive if the cold object was not there—and so the warm object is kept warmer than it otherwise would be. Radiation is not somehow preferentially attracted only to colder objects. Thus the presence of cold objects can keep warm objects warmer!!! This does not infringe the 2nd law of thermodynamics, only a schoolboy misunderstanding of it.*

Hoping to have eluded the Groundhog Day Effect that your comment was caught in, I'll remark that the warmer object will now be inclined to radiate a lot more energetically than it otherwise would have.

At lower troposphere densities it will be even more inclined to pass on the energy, through collisions, to non-absorbing species which then convect upwards.

When your feet are cold, it takes a painfully long time for them to warm up by merely putting on wool socks. If you change those socks out repeatedly for a colder pair, your toes will turn blue.

It's not hard to understand that GHG's provide the air with warmth, but it seems the ground is not really significant in the back-radiation scenario. In fact the warmer the surface becomes the more the energy will be radiated to space without bouncing around in the atmosphere.

Ryan:

> *And the problem that all these scientists have is that they will never be able to test if their theories are correct, because the time spans are too long.*

Is that so? What if we had a month off for CO_2 emissions as a globe, with CO_2 output deliberately reduced by say 30%. This would create a point of inflexion in our CO_2 output which in theory should then be followed by a point of inflexion in rising temperatures, thus proving AGW theory and demonstrating cause and effect. It would then allow us to go further and estimate the actual impact of given increases in CO_2. Got to be worth doing just as a global experiment before making any solid commitments to changing our habits.

(Well in theory, in practice I guess it would just get lost in the noise of normal weather patterns which would only prove the futility of the whole AGW nonsense.)

Dave Springer: P Wilson says:

> *...actually, water vapour above this level acts as an absorber of radiation: the so called greenhouse effect...*

Water vapor is invisible. A cloud is composed of water droplets not water vapor. Water vapor coming out of a tea kettle is invisible until it has cooled enough to condense into water droplets. This is barely high-school level physical science. You need to go back to high school and relearn this stuff and that's presuming you ever learned it in the first time through which at this point is a matter of grave doubt. You'd do miserably on the TV show *Are you smarter than a fifth grader?*

Oliver Ramsay: I didn't mean to suggest that heating the surface more would cool the air!

Just that returning energy to the surface provides more of a direct avenue for escape than if it relied only on radiation from GHG's.

P Wilson: Dave Springer says:

Largely irrelevant comments re CO_2 though correct ones regarding the Sun. However, the argument is that CO_2 is largely irrelevant to the climate, since most heat escapes via convection and conduction, hence the quick cool down at night. This cool down, and corresponding cool upper layers, show just how quickly this takes place. Nothing to do with the S-B equation. To do with observed phenomena the main reason is that there is no heat being generated, or stored by greenhouse gases, and so back radiation doesn't occur. Even if it did, the quantities would be almost infinitesimal. In your later analogy however, it is convection that is being vastly reduced in a thermal flask, which prevents quick cooling.

This convectional trapping process does not occur with outgoing IR radiation through CO_2 or any other GHG.

However, even at ground-10,000 metres, what CO_2 does is so infinitesimally small that it might as well be factored out of climate projections.

At 15microns peak, that makes not an iota of difference to

atmospheric temperatures. Besides, gases are not blackbodies, and are 3 dimensional. Radiative equations can't be applied to them (air is a very poor conductor), and so is irrelevant to climatology. With GHGs, *blocking* operates in all directions, so doesn't make any difference to the atmospheric heat content.

If you're talking about the shoulders of CO_2, then at this level there is more nitrogen and oxygen per CO_2 molecule in this area. Dilution reduces the temperature increase per unit of energy.

P Wilson: Dave Springer says: I said:

> ...*actually, water vapour above this level acts as an absorber of radiation. the so called greenhouse effect—water vapour is indeed invisible.*

The above comment was a contradiction of Izen's, who said that water vapour higher than clouds have no effect. He said:

> *Above the main low cloud layer water vapor is much less important as an atmospheric absorber/emitter.*

I think you've got it muddled somewhere, although if you wish to trace a source of the confusion between clouds and water vapour, I suggest you consult Izen's first.

George E. Smith: An interesting dialog; but perhaps KLC didn't ask the right questions, so in turn MC didn't really provide the needed answers; although what MC DID say is not incorrect.

One issue raised in the discussion was the 100 Watt light bulb, versus the 300 W/m^2 from the sky—the first one feels "hot"; while the second one doesn't. MC says that the 300 W/m^2 corresponds to -23.4 deg C. That ties in well with Trenberth's 390 W/m^2 corresponding to +15 deg C or 288 K.

But here's what MC didn't tell KLC; and we have all seen it

with that 100 W light bulb "radiating" onto two samples of dry air; one containing more CO_2 than the other. "The Science Guy" Bill Nye has performed that quite fraudulent experiment in public.

Here's what's wrong with it and what MC forgot to tell KLC.

The 100 Watt light bulb, is a tungsten filament lamp (likely gas filled), and it will emit a roughly black body like spectrum (thermal spectrum); and MC explained how near BB and real BB sources can be related via an emissivity; and of course most practical sources, are not a single temperature source so they will be somewhat spectrally discombobulated as well.

But back to that 100 Watt lamp. Which is fairly typical of the so called "warm white" light source. It will radiate a spectrum that is something like a 28-2900 deg thermal spectrum; i.e. a somewhat BB like source of radiation with a color temperature of about 2900 K. THAT IS HALF THE TEMPERATURE OF THE SUN's SURFACE.

The Sun emits a thermal (BB like) spectrum at about 64 million Watts per square metre (at the Sun's surface). Our innocent 100 Watt lamp, at half the temperature is only emitting 4 million Watts per square metre, since that goes as T^4, per S-B.

More importantly, the 100 Watt "light bulb" is ten times the temperature of the 288K Trenberth average temperature of the Earth, and the presumed source of that 390 W/m^2.

So the incandescent light bulb is ten times the average Earth temperature, and is radiating 10,000 times as much energy per square metre as the average Earth.

So now we kick in with the Wien Displacement Law that MC also mentioned.

At 288K (+15 deg C or +59 deg F), the LWIR radiation spectrum, that is emitting a total 390 W/m^2, has a spectral peak wavelength at 10.1 microns. And assuming a single temperature source, 98% of the LWIR radiation is contained between 1/2, and 8 times the peak wavelength so from 5.0 to 80 microns, is the spectral range of the average Earth temperature roughly BB spectrum.

So, colder places will radiate even longer wavelengths, but being colder, they also are radiating even less, so they don't really contribute much energy on the long wave end. For the highest surface desert temperatures which can be above +60 deg C, the spectral peak could be as low as about 8.8 microns from the Wien Law, so that will radiate down to maybe 4.4 microns, and also the total S-B emission can be 1.8-2.0 times as high as Trenberth's 390 W/m^2.

So it is those hot dry desert regions that are the principle radiative coolers of the Earth. The 8.8 micron peak is even further away from the 15 micron CO_2 band, and the peak moved further into the "atmospheric window" where water vapor is somewhat benign. Well there's little water vapor in those arid deserts anyway. But don't forget that Ozone kicks in a dip at around 9.6 microns; but it is quite narrow, because of the height, and low density and temperature of the ozone layer (less line broadening)

Now back at our phony heat lamp at 2880K; not only is it emitting 10,000 times the emittance of the average Earth; but the spectral peak is not 10.1 microns either, but is now 1.0 microns.

Well wouldn't you know it; water (H_2O) has several absorption bands near there at 0.94 microns, and also at 1.1; and supposedly humans are 95% H_2O.

So the peak radiant emission from a 100 Watt light bulb is at the correct wavelength to cause strong absorption in human flesh; no wonder it feels warm.

The 288K mean Earth surface on the other hand is radiating at 10.1 microns, and the water absorption coefficient there is about 1000 cm^{-1}; which gives a 1/e transmission depth of 10 microns; or a 99% absorption depth of 50 microns; 2/1000 inches.

It doesn't even make it through the surface of your skin; and even if it did, it wouldn't register against your body Temperature of 98.6 deg F or 37 deg C.

So no wonder the 100 Watt lamp feels warm to your skin, and no wonder it warms the air samples. CO_2 has absorption bands at around 2.7 microns, and I think also at 4.0 microns. The 2880K lamp spectrum is going to put 98% of its energy in the spectral range from 0.5 Microns (green) to 8.0 microns, so both of those CO_2 bands are going to be activated. And even the 15 micron band will absorb much more energy from the 2880K source than from the 288K source.

One should also note that although the S-B equation says the total emittance goes as the 4th power of temperature; the spectral peak emittance actually goes as the fifth power of the Temperature (T^5).

I'll leave it to the math geeks out there to figure out how much more spectral emittance you get at 15 microns, from a 2880K sources compared to a 288K source (Watts per m^2 per micron wavelength.

So like I said, the 100 Watt lamp demonstration is a total fraud. A much better source for 300-390 W/m^2 LWIR radiation that is properly spectrally peaked at about 10 microns, would be an ordinary 16 ounce bottle of water. I don't recommend the French Evian water; it's too expensive and doesn't radiate any more than any other water; use tap water it's cheaper.

A C Osborn: Oliver Ramsay says:

> In fact the warmer the surface becomes the more the energy will be radiated to space without bouncing around in the atmosphere.

WUWT?

So the surface has a direct connection to space then? Can I ask all the Physics guys a simple question?

> What happens when 2 photons travelling in opposite directions collide?

P Wilson:

@George Smith, Dave Springer maintains that GHGs—particularly

CO_2 slows the rate of cooling of the atmosphere, and uses the analogy of a thermal flask (which limits conduction and convection).

I'm arguing that CO_2 doesn't make much measurable difference to atmospheric heat content and cool rate, although, George, it depends what posts you're referring to.

Tenuc: steven mosher says:

> Yup, but there is more to it than that. If you want to do IR design today (of sensors, or counter measures or stealth technology) the physics you use is **the same physics that many skeptics deny**. good thing we don't let them design the machines that our safety as a nation depends upon.

Got out of bed the wrong side today Mosh??? Only an idiot would think that tarring all sceptics with the same brush is a sensible way to have a useful debate.

My own take on this is simple. Although lab experiments show CO_2 could raise Earth's overall temperature by a tiny amount, the IPCC's predictions of catastrophic temperature increases produced by carbon dioxide have been challenged by many scientists and found to have no substance in the real world.

The importance of water vapour is frequently overlooked by environmental activists and by the media. The large temperature increases predicted by many computer models are non-physical and inconsistent with results obtained by basic measurements—much scepticism is warranted when considering computer-generated projections of global warming when these same models cannot even predict existing observations.

P Wilson: Okay, I think I gleaned something re: the shoulders of CO_2, 15microns is the peak.

All I can fathom is that Dave Springer argues that the entire shoulders and peaks of CO_2 count, but that at the shoulders—or else the band either side around 15microns—then they become miniscule as heat absorbers. At the tropospheric level where the outer bands absorb, there are more oxygen and nitrogen competing for heat than CO_2.

Oliver Ramsay:

> *In fact the warmer the surface becomes the more the energy will be radiated to space without bouncing around in the atmosphere.*
> *WUWT?*
> *So the surface has a direct connection to space then?*

Yes.

Dave Springer: cal says:

Most of the energy the Earth receives is visible light from the Sun. The 5200K blackbody spectrum carries a significant but minor fraction of its total energy in the near infrared but no practical amount of the energy is in the far infrared where CO_2 absorption begins.

Conversely the Earth (discounting reflected visible light during the day) emits essentially no energy in the near infrared and essentially all of it in the far infrared. The amount of the energy emitted in the far infrared necessarily over time is equal to the amount of shortwave energy absorbed by the ocean during the day. Near infrared hitting the ocean surface is absorbed within the first few microns and doesn't heat the ocean but is rather carried off in water vapor which rises by convection where the energy is released when the water vapor condenses into a cloud.

However, visible light penetrates over 100 meters and is almost completely absorbed with very little of it reflected. Land surfaces to a small degree and clouds to a high degree reflect incoming visible light straight back out into space. Estimates of the

Ken Coffman

Earth's average albedo are in the 35% range plus or minus a few percentage points depending on who you ask and what they need to stick into their climate models to better reproduce paleo-climate data. Albedo is used as a fudge factor to tune climate models for better fitting climate hind-casts.

So anyhow, about 60% of the solar energy arriving at the top of the atmosphere works to heat the ocean and all that energy eventually radiates out at night at much lower frequency. It's the difference in CO_2 frequency response to short wave versus long wave radiation that allows it to act as an insulator. It's transparent to short wave radiation and opaque at a few significant long wave frequencies so almost all solar energy passes through to the ocean unimpeded during the day but impedes long wave energy emitted by the surface at night.

The simplest and quite accurate way to conceptualize this is that CO_2 is an insulator—it's like a blanket over a dark rock where you remove the blanket during the day so the rock can heat up in the Sun and you put the blanket back at night to retain more of the warmth from the Sun during the night. The end result of blanket vs. no blanket is a warmer rock with the blanket than the rock would be without it.

Ian W: When water vapor in the atmosphere condenses into liquid water and then changes state again and becomes ice, it gives off latent heat for both state changes.

Does that latent heat release follow Stefan-Boltzmann's radiative equation?

Mikael Cronholm says:

No Ian, latent heat is not really related to S-B.

So if you consider all the clouds and it is estimated that 62% of Earth's surface is covered by clouds—the amount of energy released as latent heat is huge. Yet people persist in working out the amount of heat leaving the surface using Stefan-Boltzmann radiation equations. This is obviously incorrect.

izen:

> *Some energy leaves the SURFACE by convection, but it does not get very far.*
> *It is not clouds that block it—strange idea!*
> *It's the adiabatic lapse rate.*

Convection can only move air until the increased buoyancy from the lower number of molecules per cubic metre is offset by the lower density with increasing altitude. It is the lower density of the atmosphere with altitude that blocks convection. The low altitude of most clouds indicates that the temperature falls below freezing just a few Km above ground level and the atmosphere regains that latent heat of evaporation well below the Tropopause. Above the main low cloud layer water vapor is much less important as an atmospheric absorber/emitter.

Well I don't know about 'not getting very far' 60,000ft or more in the tropics is the level of the Tropopause (the top of the convective atmosphere) and in the Inter-Tropical Convergence Zone (ITCZ) the cloud tops of towering storms are (by definition) up to that level. (A recent Northwest Airlines flight experienced liquid rain hitting the aircraft in cruise above 30,000ft in a storm between Hong Kong and Tokyo despite the ambient outside-air-temperature being well below zero. The rain froze into ice on hitting the airframe). The height of the cloud tops all depends on why the clouds are there.

Humid air is more buoyant than dry air, so just being more humid is enough for a volume of air to start convection. Drier air sucked in will then pass over the wet surface and Henry's law applies and more water will evaporate into the dry air cooling the surface. No external heat source is required. As the humid air rises the air temperature will drop at the wet adiabatic lapse rate and water will start to condense around nucleation particles and form clouds. This type of cloud with a gentle wind will become the low stratocumulus over oceans and coastlines. And each and every cloud droplet has taken heat from the surface and radiated it as it becomes a droplet and again when it becomes ice.

With the application of tropical heat and humidity convective storms develop every day some towering ten miles high into the atmosphere. If the conditions are conducive these storms merge and the Coriolis force on the air being rapidly drawn in at the base of the storms starts the winds in the storm system to rotate. There is a lot of discussion of the ACE index here, but to put it in perspective, the amount of energy released by a hurricane can be calculated based on the latent heat emitted and/or the kinetic energy. An average hurricane in a day transfers latent heat equivalent to 200 times the world-wide electrical generating capacity.[130]

As confirmed in response to my question at the top of this thread, the temperature of the cloud has no bearing on the amount of latent heat being released. The heat being released is linked to the amount of water changing state and the latent heat of that state change.

The missing heat is almost certainly in the misunderstood and underestimated hydrologic cycle that is ignored in favor of a simple Stefan-Boltzmann radiation equation.

Jim D: Slacko, following up on the reply already by cal, solar radiation contains a small fraction of IR, maybe only a few W/m^2, so most of the warmth you feel is from absorbing the visible/UV rays.

The other part of Slacko's question: yes, N_2 can't radiate IR, nor can O_2 or argon. This is from physics.

Robert Clemenzi, yes, $324 \ W/m^2$ corresponds to 34°F. This is supposed to be a global and annual average value, which obviously varies greatly locally. I think your suggestion is that it implies reflection because the clear atmosphere is probably colder than this. Actually clouds behave as almost perfect black bodies, so the cloud bases are emitting at their temperature, which may account for the

---

[130] http://www.aoml.noaa.gov/hrd/tcfaq/D7.html

average being as high as 324 W/m^2. In the tropics even clear skies can emit more than 324 W/m^2 to the surface because of the high H2O contents.

George E. Smith:

> ...at 15microns peak, that makes not an iota of difference to atmospheric temperatures. Besides, gases are not blackbodies, and are 3 dimensional. Radiative equations can't be applied to them (air has a very poor conductor), and so is irrelevant to climatology. With GHGs, 'blocking' operates in all directions, so doesn't make any difference to the atmospheric heat content...

Either P. Wilson or Dave Springer said the above; it contains both of their names at the top. I'm inclined to discount Dave as the source of that statement; but can't tell where if anywhere Dave stops, and P. starts.

In any case, while it may be pedantically true that gases aren't black bodies; we know that since absolutely nothing real is a black body. But where in the Physics texts does it say that water or ice stops radiating thermal radiation, the instant that it changes from a liquid or solid, into a gas at the same temperature. Has anybody ever recorded a motion picture video of that shut-off process actually taking place in any Physical system.

Can somebody cite some peer reviewed literature that forgives gases from radiating thermal radiation; that is Electromagnetic Radiation of a spectral nature that depends on the temperature of the gas. And where in that peer reviewed paper, did it say that the Sun is further excused from obeying the non thermal EM radiation prohibition of gases?

One thing we can be fairly sure of, is that the Rayleigh scattering of short wavelength sunlight, which makes the entire daytime sky appear "sky blue" when viewed in any direction (not counting near the Sun); and that means it looks the same looking up as looking down; because the large angle scattering due to the RS process makes the atmosphere an isotropic source of sky blue light which originally came from the Sun.

By the same token, the LWIR emissions from the atmosphere; whatever their source; are also isotropic, since there is no preferred direction of emission. So the infrared sky looking up towards space is pretty much the same as looking down from outer space; except (apparently) that according to Trenberth, 40 W/m^2 of LWIR radiation that is actually emitted from the Earth's surface, actually escapes unharmed to space, and the other 350 W/m^2 is absorbed by the atmosphere. Apparently per P.Wilson, it cannot be subsequently emitted from the atmosphere per the radiation laws governing BBs and other thermal radiation laws; which he says don't apply to gases.

So now would somebody not as dense as I am, like to explain to me what is the source of quite thermal spectrum looking (grey body) radiation that is seen from outer space, when looking at the Earth. Only 40 W/m^2 of it can be coming from the ground per Trenberth and it does not have a narrow CO_2 absorption band spectrum as one would expect, if direct emission of molecular resonance radiation spectra from CO_2 was the source.

Several have noted that a lot of "heat" transport to the upper atmosphere is a result of convection of hotter surface gases into the upper regions. Somebody could explain for us how that eventually escapes to space, seeing as how gases can't radiate thermal spectra according to the black body or S-B radiation laws.

Clearly the ten years, that I formally studied Physics in school, plus the subsequent 50 years of using it daily in practice to accomplish things; was not enough for me to come across that jewel of knowledge, that gases do not radiate according to the radiation laws.

Readers might also find it interesting to look at Figure 11; *Higher Members of the Balmer Series of the H Atom (in Emission) Starting from the Seventh Line, and showing the Continuum (Hertzberg41)* The reference to "hertzberg 41" is the G. Herzberg Ann. Physik (4)

84, 565 1927 citation

And that photo of an actual real measured, scientific observation of a (non teracomputer simulation of) Balmer spectrum including a "Continuum Spectrum" along with the Balmer line spectrum is in *Atomic Spectra and Atomic Structure* by Gerhard Herzberg; Prentice Hall 1937. Herzberg mentions that in Emission the continuum corresponds to a free electron of any energy being captured by a proton, and going in to "the orbit" having the principal quantum number n=2 (this is of course in the Bohr Atom interpretation. Modern Quantum formulations may be different.)

And nowhere does temperature get mentioned here. This continuum radiation still corresponds to an energy level transition; in this case from an ionised state. Well of course you won't find any ionised states on the Sun I would imagine; that spectrum is commonly observed in stars. But the end of the Balmer series, is at about 3800 Angstroms; so that continuum is in the UV region; not in the LWIR region where the thermally originated LWIR emission spectrum occurs.

The extraterrestrial LWIR spectrum from the Earth shows a narrow spectral dip at the 9.6 micron Ozone band, and a wider one at the 15 micron CO_2 band from about 13.5 to 16.5 microns, otherwise it looks pretty much like any other near B-B spectrum as limited by the Planck and S-B laws (Wien also).

barn E. rubble: The question posed early on this thread from KLC:

> From your point of view as an IR expert, does this aspect of the global warming theory make any sense?"
> RE: ". . . notorious Ternberth/Keihl energy balance schematic (as shown in Figure 1 of this paper: http://www.cgd.ucar.edu/cas/Trenberth/trenberth. papers/TFK_bams09.pdf), you see the back radiation is determined to be very significant...

Maybe I missed it but was there a summary or (for want of a better term) a consensus from the posters here that can answer this with some authority? MC replied he had issues with the numbers used. I was wondering what others thought re: reasonable numbers if those referred to (Fig 1) are not reasonable.

MC also noted, "Every question answered raises a few more, which grows the confusion exponentially."

I was hoping to clear up some confusion on my part.

Jim D: barnErubble, MC was not concerned with the main number of > 300 W/m^2 from IR as back radiation in those papers. He was concerned with the accuracy of the residual 0.9 W/m^2, and how anyone could accurately state such residuals given the large canceling numbers involved. This is rather related to Trenberth's "missing energy" issue, and is a valid concern.

P Wilson: George E. Smith says:

> *Clearly the ten years, that I formally studied Physics in school, plus the subsequent 50 years of using it daily in practice to accomplish things; was not enough for me to come across that jewel of knowledge, that gases do not radiate according to the radiation laws.*

Of course they do, as only you would know. I'm referring to the S-B equation in particular as applied to CO_2, and the fanciful results on paper that it generates. I think you've implied before that except for a tiny fingerprint of black body radiation, most radiation evades CO_2 in its escape to space.

Before energy escapes into space through radiation, it doesn't matter whether the energy is in contact with CO_2, water vapor or nitrogen.

Robert Clemenzi: Jim D says:

...solar radiation contains a small fraction of IR, maybe only a few W/m^2, so most of the warmth you feel is from absorbing the visible/UV rays.

Actually, more than 50% of the solar energy at the top of the atmosphere is IR. At the surface, the percent is increased because some of the UV was absorbed at a higher level. On a cloudy day you can get a sunburn. However, since you will not feel a temperature difference between overcast light and full shade, the "heat" you feel on a sunny day will not be from the UV.

Sensor operator: Steeptown mentioned:

...it is evident to me that the radiative effects of CO_2 in the atmosphere are of 2nd or 3rd order compared to the radiative, convective and latent heat effects of H_2O.

Well, when do 2nd order effects become important? The problem with increasing CO_2 in the atmosphere is the ever increasing 2nd order effect. So a little bit of energy doesn't escape. No big problem. But, if the effect is compounded, which appears to be the case for CO_2, suddenly a little bit of energy becomes a lot of energy. When do we care?

Water in the atmosphere has two very different properties versus CO_2. First, the amount of water in the atmosphere is not likely to double unless there is a major change in the atmosphere. So even if H_2O is a first order effect, the much smaller changes in the amount of water is not going to be significant. However, CO_2 is increasing fairly quickly. And a doubling or possibly tripling of the amount of CO_2 is very real so the second order impact is likely to be much larger than the small deviations of the first order impact. Second, CO_2 is throughout the entire atmosphere. Sure, it is only 0.04% of the atmosphere. But if we look at the entire atmosphere, not just the surface, water is only 0.4%. Doubling or tripling CO_2 is now ~0.1 %. Suddenly CO_2 is not that small after all.

Something else folks seem to be forgetting with CO_2 is the bands it absorbs. In particular, it is absorbing in bands that water is not. At lower levels of CO_2, a sufficient amount of LW radiation was escaping to provide an energy balance. With the increased amount of CO_2, the balance has been upset and the Earth is trying to reach a new equilibrium.

Some folks are very taken with Mr. Cronholm's one comment:

> For a theory to be scientifically proven, it has to be stipulated and tested, and the test must be repeatable and give the same results in successive tests for the theory to be proven.

This may be true, but we only have one planet. We don't have an alternate Earth that we can play with to see what happens. Also, tested does not necessarily mean run in a lab or a model. Consider evolution. We don't "test" evolution, we look at the available data and the composite result from this evidence is the fundamental reason for evolution. Climate change is similar: we look at temperature, CO_2, extended growing seasons, northern and southern boundaries of migrating birds and insects, extreme weather events, receding glaciers, etc.

One comment that has been discussed time and again is that in the geological record, an increase in CO_2 occurs after an increase in temperature. Fine, let's assume that is true. The problem we have is the measurements we are making now show CO_2 increasing and temperature following. What does it mean? Simple: a different mechanism is now responsible, not the natural processes we have been able to identify from past changes.

Of course, there could be another natural phenomenon we have not found/discovered. But as Dr. Alley from Penn State has pointed out: we built the satellites, we made the measurements, when someone disagreed we made new satellites and new measurements. We have spent billions of dollars and so far the overwhelming evidence is in support of AGW.

Think of it like Children's Hospital. The doctors there have one job: to put themselves out of a job. If they could solve all the medical issues they see, they would no longer be needed. And that is actually their goal! They don't want kids to suffer. Do people really think climate change scientists want AGW to be true? If they are right, the end result is really bad.

Don V: AC Osborne: re photon-photon collision
I am not a physicist but, back when I went to college and took physics, I was taught that photon-photon collisions are not possible.

Being virtually massless and only possessing momentum, two photons can only indirectly interact with each other. Their interaction is called Delbrück scattering. When they "interact" they annihilate into a virtual electron-positron pair, which then annihilates back into two real photons again. They lose no momentum or energy in the process.

Don't know if that is completely right, my memory is hazy, but my rule of thumb has always been that for first approximations photons are massless and that massless quanta's of energy basically just go right thru each other.

cba: I found the explanations both fundamentally accurate to my understanding level and very nicely presented in what should be a very understandable way.

One major error though is the misconception that $0.9 \text{ W}/\text{m}^2$ imbalance is measured. It is a model calculation based upon plenty of presumptions. The references to the CERES & ERBE data actually turn up to have accuracy problems of several W/m^2 so despite all the discussions in the papers, they resort to modeling to get their value. In the earlier papers, such as Hansen's, the uncertainties in the modeling for such things as cloud formation versus T are discussed and decisions made that gives rise to their claim that cloud formation decreases with temperature which is not the case for all presumption options like cloud formation depending upon absolute humidity levels.

Barn, the details isn't so much radiative theory as it is to the system of Earth and its atmosphere. You can rather safely bet that the warmer crowd will minimize every number that limits AGW and exaggerate every number that supports it.

Primary amongst the real uncertainties is that of the albedo and cloud cover and the behavior of the cloud cover with conditions such as temperature. The whole premise being used that there is one surface temperature average for a given blockage of outgoing IR is totally flawed. While incoming solar average power is given great importance, the notion of what the albedo reflects away is often considered to be constant and not the actual variable it really is, driven by many other factors. If you consider the comments of the warmers, they give lip service to land use changes and loss of snow and ice cover as being the important factors with albedo. Funny how about how 80% + of the albedo is due to 60% + cloud cover and of that under 20% surface contribution, around 70% is oceans and that leaves next to nothing for contributions from the land surfaces, never mind ice and snow and a little rain forest or two.

K&T97 admit possible errors of up to 20% in their cartoon values. Amazing how close they got to that by underestimating cloud cover and overestimating land and surface contribution. It actually looks like they forgot to take cloud cover into account and gave a clear sky only value. Later they use the same values and apparently forgot to include their serious margin of error.

izen: Ian W says:

> As confirmed in response to my question at the top of this thread, the temperature of the cloud has no bearing on the amount of latent heat being released. The heat being released is linked to the amount of water changing state and the latent heat of that state change.
>
> The missing heat is almost certainly in the misunderstood and underestimated hydrologic cycle that is ignored in favor of a simple Stefan-Boltzmann radiation equation.

The *Science of Doom* site[131] has a good discussion of some of these issues.

The amount of energy transported by latent heat changes is complex; there is one simple way to quantify it however. The total global yearly rainfall is a direct measure of how much energy was moved by evaporation and condensation in the hydrological cycle.

But the first LoT kicks in. The energy is still around. The only way it can get of the planet is by radiating IR photons into space. All the hydrological cycle can do is move the location of that emission around. That evens out the temperature so that water does not boil at the equator and can melt at the poles, but the same amount of energy has to leave the planet. Otherwise it will warm until the S-B T^4 relationship increase the emissions enough to compensate.

Jim Masterson: Smoking Frog says:

> The average of the minimum and maximum overlaps gives exactly 62%:
> $((49 + 6 + 20) + 49) / 2 = 62$, but I'm not sure that this is what we'd get with random overlap, even if, as my calculation assumes, there are no real-world constraints.
> Interesting. It may be what the authors were doing.
>
> The Inclusion-Exclusion Principle requires that we know the overlap(s) to begin with, so it can't give us the answer.

But it does give us an answer that is almost the same:
$.49 + .06 + .20 - (.49)*(.06) - (.06)*(.20) - (.49)*(.20) + (.49)*(.06)*(.20) = 0.61648$.

Oliver Ramsay: izen says:

> But the first LoT kicks in. The energy is still around. The only

---

way it can get of the planet is by radiating IR photons into space. All the hydrological cycle can do is move the location of that emission around.

You make it sound like water just shuffles energy around horizontally.

This is claiming the taxi delivers you to your holiday hotel door, not the airplane.

cba: Frog, jim, the details show in kt97 that the assumptions are 100% optical thickness for the two main cloud types and 50 or 60% thickness for the 6% coverage contributor. The 62% result they come up with is assuming 100% thickness with random overlap. It does seem to agree well with what little is known of the actual coverage, which can vary substantially over time, something like over ±5% as I recall. That also leads to an albedo variation of around 5% and a peak to peak difference in reflected incoming light power of about 10 w/m^2, far more than a mere CO_2 doubling.

JAE: Hans Erren says:

> *Adding water vapour in the tropics gives the tropics a greenhouse effect, I remember very well when I left the air conditioned hotel in Dhaka Bangladesh, a wall of humid heat hit me and I had to wipe my glasses.*
>
> *Every added particle to the atmosphere is a small downward radiator, adding more particles therefore adds more radiators. That's in a nutshell the greenhouse effect.*

Hmmm. Then why is the maximum temperature there NEVER as hot as it is almost every day from June-Sept. in Phoenix, AZ?

George E. Smith: Ian W says:

When water vapor in the atmosphere condenses into liquid water and then changes state again and becomes ice, it gives off latent heat for both state changes.

> *Does that latent heat release follow Stefan-Boltzmann's radiative equation?*

Latent heat and anything to do with heat, is a property of physical matter; it has nothing to do with Electromagnetic Radiation; so it has nothing to do with the Stefan-Boltzmann equation.

EM radiation can go anywhere it wants to; heat can't.

Matter: Interesting article and Al Tekhasski covers a good point too.

Detectors are tuned to work in a certain wavelength. We can make materials that have energy levels (or bands) that only allow transitions of certain energy. One common way of doing it is to create a 'quantum well' (look it up!) using sandwiched layers of different materials, but there are others.

The atmosphere is largely transparent in some wavelengths, it doesn't absorb well here. This means that there aren't many available transitions, so it can't emit here well either. So if you point your scanner upwards and measure, and your scanner is tuned to this energy band then you won't measure many photons coming down. This will mean a small electrical signal (the voltage output is related to the number of photons) and your device thinks 'small voltage, so not many photons, so what I'm measuring is cold'.

This works because most objects have emissivity close to 1 at these ranges, whilst the atmosphere doesn't. So you can measure the photons coming from far away objects and not the atmosphere; if the atmosphere had emissivity 1 for these wavelengths then it would absorb the light from the object before it could reach your scanner, and you would always measure the atmospheric temperature—not useful for a long range thermometer!

The strength of CO_2 absorption can be tested empirically by spectral measurements from satellites (and changes can be measured from ground stations) for a variety of conditions. Look up some of Philipona's papers or the Harries 2001 paper. They provide experimental confirmation of the physics.

Latitude: physics schymisics, real world...

When CO_2 levels were 4000PPM, we had an ice age. When CO_2 levels were 3000PPM, we had an ice age. When CO_2 levels were 2000PPM, we had an ice age and here we are wringing our hands and wetting the bed over 390PPM.

Real world says the only tipping point is when CO_2 levels get in the thousands, we can have another ice age...

Jim D: I will correct my answer to Slacko, taking into account Robert Clemenzi's point.

Yes the IR at night is mostly due to H_2O and CO_2. During the day, the Sun produces near-IR in the 1-2 micron range, while the atmosphere only emits at the wavelengths longer than about 4 microns. These two types of IR may be of comparable magnitude in the upper atmosphere, but much of the near-IR is absorbed before reaching the surface.

kuhnkat: Sensor operator says:

> *Think of it like Children's Hospital. The doctors there have one job: to put themselves out of a job. If they could solve all the medical issues they see, they would no longer be needed. And that is actually their goal! They don't want kids to suffer. Do people really think climate change scientists want AGW to be true? If they are right, the end result is really bad.*

Sensor, I don't have to THINK that is what some climate scientists think. Phil Jones actually stated it in his emails.

Bill Illis: How does Nitrogen and Oxygen warm up and cool down in the atmosphere?

The way it is described by some here, the radiation theory assumption is that both of these gases are at absolute zero.

kuhnkat: Don V says: AC Osborne: re photon-photon collision.

Wikipedia has an interesting article on photon-photon collisions.[132]

Of course it points out that photons can't collide since they don't become photons, or quantized, until they interact with matter. This gets to the root of the issues with backradiation. Until the radiation from the CO_2, or the Earth for that matter, actually interacts with a particle they are best described through wave mechanics which was well mapped.[133][134]

Many people ask what would happen to the energy if 2 waves cancelled.[135]

Sometimes the most simple concepts are missed. Actually it may not be simple reflection but scatter, although the scatter would seem to be when less than 100% cancellation occurs.

George E. Smith: Bill Illis says:

> How does Nitrogen and Oxygen warm up and cool down in the atmosphere. The way it is described by some here, the radiation theory assumption is that both of these gases are at absolute zero.

Funny you should notice that too, Bill. N_2 and O_2 at 288 or 300K may not seem to be emitting any thermal radiation; but that is because we are used to feeling our "heat" at much shorter wavelengths, and much higher temperatures.

---

[132] http://en.wikipedia.org/wiki/Two-photon_physics
[133] http://en.wikipedia.org/wiki/Phase_cancellation
[134] http://www.mrelativity.net/Papers/4/Rykov.htm
[135]

http://newsgroups.derkeiler.com/Archive/Rec/rec.radio.amateur.ante nna/2005-12/msg00243.html

Considering how difficult it is to observe and measure thermal radiation at 288K; especially spectrally resolved; it is no wonder that ordinary humans are not even aware of its existence. We certainly can't feel it on our skin.

But shift the wavelength down by ten, and raise the temperature by ten, and the Total Radiant emittance by 10,000 times, and the spectral peak emittance by a factor of 100,000 and humans finally can be made aware of its presence.

But I don't see a whole lot of 100 Watt "heat lamps" turned up skywards, anywhere I've ever been!

Jim D: Bill Illis, nitrogen and oxygen heat and cool by conduction and convection only, not by radiation.

George E. Smith: Hans Erren says:

> *Adding water vapour in the tropics gives the tropics a greenhouse effect, I remember very well when I left the airconditioned hotel in Dhaka Bangladesh, a wall of humid heat hit me and I had to wipe my glasses.*
>
> *Every added particle to the atmosphere is a small downward radiator, adding more particles therefore adds more radiators. That's in a nutshell the greenhouse effect.*

So what is the magic of these particles that they know to only radiate downwards? Common sense would say that whatever in the atmosphere is emitting radiation of any kind, is doing so pretty much isotropically with no directional bias.

So whatever your particles are emitting downwards, they must be emitting a like amount upwards; which escapes to space. So whatever the original source of the energy that your particles are radiating; be it direct incoming sunlight or surface emitted LWIR radiation, it seems that half of it is going to escape to space, and not reach the ground. Particularly when the Sun is the source of that energy your particles are radiating downwards; that is an amount of sunlight that will never reach the ground; so it will get less hot than if your particles did not intercept that solar radiation.

No matter how you try to skin the cat; anything in the atmosphere that absorbs any incoming solar energy or even widely scatters it, such as the blue skylight due to Rayleigh scattering, must result in less solar energy reaching the surface of the Earth (ocean) and getting stored in Earth's thermal sink.

Ian W: izen says:

> As confirmed in response to my question at the top of this thread, the temperature of the cloud has no bearing on the amount of latent heat being released. The heat being released is linked to the amount of water changing state and the latent heat of that state change.
>
> The missing heat is almost certainly in the misunderstood and underestimated hydrologic cycle that is ignored in favor of a simple Stefan-Boltzmann radiation equation.
>
> The 'Science of Doom' site linked to from here (top right of page) has a good discussion of some of these issues. The amount of energy transported by latent heat changes is complex, there is one simple way to quantify it however. The total global yearly rainfall is a direct measure of how much energy was moved by evaporation and condensation in the hydrological cycle.
>
> But the first LoT kicks in. The energy is still around. The only way it can get of the planet is by radiating IR photons into space. All the hydrological cycle can do is move the location of that emission around. That evens out the temperature so that water does not boil at the equator and can melt at the poles, but the same amount of energy has to leave the planet. Otherwise it will warm until the S-B T^4 relationship increase the emissions enough to compensate.

So when the heat is released at 30,000 feet[136] in the atmosphere as liquid water turns to ice, well past the dense part of the atmosphere it does not radiate to space?

How does it know that it can only radiate down? ;-)

It certainly appears that significant IR is being emitted from these clouds and weather systems independent of their temperature.[137]

cba: Bill Illis, N_2 and O_2 are at the same temperature in a small parcel of air, just like the CO_2. The CO_2 absorbs some energy and the average time it takes to radiate it away is more than enough time for it to be transferred away by collisions, most likely N_2 followed by O_2. Just because a CO_2 molecule excited by a photon is likely to have its energy reduced by a collision, so too is the likelihood that a CO_2 molecule will be excited by a collision and capable of emitting a photon or capable of being 'defused' by yet another collision.

The net result is that a certain fraction of these molecules will emit a photon of a particular energy (wavelength) based upon the temperature and upon the proclivity of the CO_2 molecule to absorb or emit that energy. Absorption doesn't really depend much on temperature but the emission is highly dependent. The blackbody curve for a given temperature is actually a portrayal of the fraction of molecules at particular energies that are capable of emitting photons. The BB curve must be a solid or liquid—or an optically thick enough gas to be thick at all wavelengths. For the Sun to essentially emit a 5800K BB curve from the photosphere, one has many heavier elements present that are ionized and so are capable of emitting a continuum rather than merely a spectrum.

Mikael Cronholm: Many things depend on what definition is

---

[136] The updrafts in the ITCZ as aircraft have found out to their cost can be in the order of 100KnH or more and liquid water can get that high before freezing—hence my quote of the Northwest Airlines Airbus.

[137] http://www.ssd.noaa.gov/goes/east/natl/flash-rb.html

used for heat, but the one I stick to is that heat is the total kinetic energy of the particles in a substance (mass times velocity squared over two). Then infrared is not heat, but it is caused by, and can cause, heat. Emission creates radiation energy by converting heat energy to electromagnetic waves (or photons, if you prefer) and absorption converts in the opposite way.

Thermal radiation, if you define it as radiation that can heat or cool a surface by exchange of radiation, will be wavelengths from somewhere in the UV throughout visible and IR. Most of the heat the Earth receives from the Sun is in fact in the visible band. In the shorter wavelengths, X-ray and gamma, most objects will transmit most of it, and hence no transfer of energy. Longer wavelengths, microwave, and radio wave, do not significantly heat things, I am not sure why, but if they did heat us we would be cooking with all the radio waves around. Microwaves can only heat by directly agitating water molecules, because they are dipoles, so it is a different process than we normally think of as heat transfer by radiation. But it is a matter of definition, largely.

Normally, the shorter the wavelength the better the penetration. If you consider human skin, UV, visible and IR penetrate in that order, UV the most, visible a little bit, and IR hardly at all.

The reason a remote control does not feel hot is merely a question of magnitude. A couple of AA batteries don't have a lot of energy to give off. By the way, you can use a normal CCD camera to check if your remote works, a simple one in a cell phone will do fine. CCD's work up to around 1 micrometer.

enough: One question: I am looking for a good answer.

At sea level, when an IR photon is emitted from Earth in an upward direction, what is the distribution of its path length before it excites (is absorbed) by a CO_2 molecule and is thermalized into the atmosphere. I believe I know the answer, but looking for independent confirmation.

Harold Pierce Jr: davaidmhoffer says:

CO_2 is reasonably well mixed throughout the troposphere...

This is not true and is just flat out wrong. In real air, there is no uniform distribution of mass of the atmosphere in space and time as shown by weather maps. High pressure cells have more regional mass and dry air than do low pressure cells with moist air. And these are constantly moving sometimes quite rapidly. Humidity lowers the density of dry air by as much as 5%.

Tropical air at ca 30 deg C and with 100% humidity has 80% of the mass per unit volume than does dry cold air at STP (0 deg C and 1 ATM pressure).

Comprised of Nitrogen, Oxygen, the inert gases, which are the fixed gases, and CO_2, purified dry air (PDA) at STP has presently 390 ml, 17.4 millimoles, 766 mg or 0.000766 kg of CO_2 per cubic meter and has a density of 1.2929 kg per cu per meter. PDA does not occur in the Earth's atmosphere. The composition of PDA (i.e., relative amounts of the fixed gases and CO_2) is fairly uniform thru out the Earth's atmosphere and is independent of site, temperature, pressure, humidity which includes water vapor and clouds except for minor local variation in particular with respect to CO_2.

If PDA is cooled to -53 deg, the amount of CO_2 is 21.6 mmole cubic meter and concentration is 390PPMv. If PDA is heated to 45 deg C, the concentration is still 390PPMv but there are only 12.5 mmoles of CO_2 per cubic meter.

GCM calculations generally use the concentration of CO_2 in PPMv which is the incorrect metric and thus are fatally flawed. The correct metric is mass (or millimoles) of CO_2 per unit volume.

The water droplets of clouds contain CO_2 which can be released if they dissipate or transport CO_2 to the surface if they turn into raindrops. How much CO_2 is sequestered in the clouds? Probably a lot.

cba: enough says:

One question: I am looking for a good answer.

There's no real good answer, I'm afraid. Depending on the wavelength of the photon, it might travel from surface to space without a hint of a capture OR it might not make it 2 centimeters without a sure thing capture. Most of the critical wavelength bands fall somewhere in between. At lower altitudes (higher pressures) the individual lines are spread out essentially forming bands. At higher altitudes and lower pressures, the lines become much sharper taller and narrower. Molecules here are less likely to absorb or emit a photon whose wavelength is further from the peak.

Mikael, that's a very restricted definition of heat energy. Energy can add velocity to a molecule or it can increase the internal energy state. IR is merely an electromagnetic form of energy. The same goes for radio waves, light, IR, UV, X-rays and gamma rays.

eadler: If the mean error is known, the readings can be adjusted. If the mean error remains constant, than the temperature anomaly, which is what is being sought will not suffer in accuracy.

And those "if's" can be significant. Much of the mean error "adjustment" ability depends on the sensor, electronics, environment, calibration protocols, and how well these protocols are ACTUALLY followed.

Calibration error doesn't affect the temperature anomaly if the error remains constant. In the case of equipment changes, so that the temperature trend becomes discontinuous, the change if detected is corrected for by the use of adjacent location data that is consistent. These adjustments are normally done by using computer programs. Calibration of the equipment is not really necessary. The new database, which is the subject of this thread, will be corrected for equipment discontinuities in the same way as the previous thermometer data bases have done it.

This is getting off topic of this thread. There are other threads on this web site that deal with corrections of the temperature record, including the new data base being developed by Muller.

enough:

> One question I am looking for a good answer.
>
> There's no real good answer I'm afraid. Depending on the wavelength of the photon, it...

Sorry, question was not to the point. If an IR photon in the CO_2 band is emitted at sea level, what is its path length before being absorbed by another CO_2 molecule.

Bill Illis: Jim D says:

> Bill Illis, Nitrogen and Oxygen heat and cool by conduction and convection only, not by radiation.

Accordingly to your theory, these are meaningless concepts. They play no part in the greenhouse effect. Why do you bring them up now?

Mikael Cronholm: @ cba. Yes, I know it is a restricted definition, which I kind of pointed out, implicitly. I just thought a simplified answer would be good enough, since the main point I was making was that radiation in itself is normally not defined as heat. But there are two definitions of heat, differing only in the semantics.

cba: Mikael, heat is energy in transit. Loosely, radiation readily fits the definition. One too easily tends to forget that there is more to energy states than kinetic energy of the motion of individual molecules.

jae: N_2 and O_2 are at the same temperature in a small parcel of air, just like the CO_2. The CO_2 absorbs some energy and the average time it takes to radiate it away is more than enough time for it to be transferred away by collisions, most likely N_2 followed by O_2.

Just because a CO_2 molecule excited by a photon is likely to have its energy reduced by a collision, so too is the likelihood that a CO_2 molecule will be excited by a collision and capable of emitting a photon or capable of being 'defused' by yet another collision.

The net result is that a certain fraction of these molecules will emit a photon of a particular energy (wavelength) based upon the temperature and upon the proclivity of the CO_2 molecule to absorb or emit that energy. Absorption doesn't really depend much on temperature but the emission is highly dependent. The blackbody curve for a give temperature is actually a portrayal of the fraction of molecules at particular energies that are capable of emitting photons.

The blackbody curve must be a solid or liquid—or an optically thick enough gas to be thick at all wavelengths. For the Sun to essentially emit a 5800K blackbody curve from the photosphere, one has many heavier elements present that are ionized and so are capable of emitting a continuum rather than merely a spectrum.

I think this is all correct.

But, I would like to posit another "world" for the warmistas. Suppose the atmosphere of Earth consisted of ONLY N_2 and O_2. Now, these gases could not cool by IR emissions, but they would warm by conduction from the surface. And they could not radiate to space. So would they continually warm for millions of years?? Would they melt the planet? WTF, folks?

Jim D: Bill Illis, you are saying conduction and convection are meaningless when they explain the temperature profile of the atmosphere and a large part of the heat flux from the ground. They are vital to any complete theory. Some people have trouble with the concepts of conduction, convection and radiation all being important at the same time, I guess.

Mikael Cronholm: @ cba. Yes, that is the other definition. Both are equally valid, but when I teach heat transfer and IR temperature measurement to engineers, I find the definition that implies that heat is contained in objects to be easier to work with. I accept both definitions and the systems of nomenclature that they create; I just prefer one because of simplicity. And following this discussion it seems to me that the definition I prefer for practical reasons also seems to be the one that people use here.

If I define heat as being the thermal energy flow I run into problems explaining heat capacity for example. The term "heat transfer" becomes a bit funny if "heat" in itself is a transfer, a bit like saying "free gift".[138]

marky48[139]: Great experts for this place. An electrical engineer and a business/economics major. And these musings disprove global warming and the properties of CO_2? LOL. Check any temperature charts lately? Any Mars missions planned from your garage? How about no Watts in the bulb?

Mikael Cronholm: @ marky48. Great contribution! Very informative! Naah, not really...

When you find something interesting to post that questions or discusses the facts presented instead of arguing ad hominem and attempting to ridicule me, you can try again.

In absence of degrees on the subject, I have worked with and studied IR thermography as it is applied in industry and research for over 20 years, and I have been teaching it for 15 years or so, and written books about it. This discussion is on the fringes of my expertise, and I don't claim to be a climate scientist or expert at all, but I am quite enjoying learning about it here. And if you look a little carefully you will find that I am on neither side in the debate.

---

[138] http://en.wikipedia.org/wiki/Heat#Semantic_misconceptions
[139] KLC: I left this comment in for comic value. Marky48 combines a feeble intellect with strong, pedantic opinions in an extremely unpleasant combination.

You have apparently taken sides though, based on what, I don't know. If you would attempt to enlighten me you will have to show a little more intelligence than you just did.

Oliver Ramsay: jae says:

> Suppose the atmosphere of Earth consisted of ONLY N_2 and O_2. Now, these gases could not cool by IR emissions, but they would warm by conduction from the surface. And they could not radiate to space. So would they continually warm for millions of years?? Would they melt the planet? WTF, folks?

jae, there was a good thread on this several months ago. Although I'd like to track it down for you, I think somebody will recall it better than I do and save me the trouble.

Obviously, the surface could still radiate and accept heat from the atmosphere through collisions. Convection would arise.

marky48:

> For a theory to be scientifically proven, it has to be stipulated and tested, and the test must be repeatable and give the same results in successive tests for the theory to be proven.

Try   http://en.wikipedia.org/wiki/John_Tyndall   and http://en.wikipedia.org/wiki/Svante_Arrhenius.

Late to the party as usual, sport.

Then try these nuggets on for size. Only fools and not even oil companies dispute man's contribution to the greenhouse effect and the observed effects of global warming. If you are smarter than NASA you haven't shown it here. Get out or lose your reputation. If you value the one you claim to have, that is.

> The question is whether a minor trace gas controls the climate. If you believe you have solid evidence that it does, provide it.

Ken Coffman

—Smokey

See the above, smokestack. Ever heard of exponents? It sounds like you've had too much of the emissions already though like the rest of these toasties.

Oliver Ramsay: @ cba
That CO_2 imparts warmth to N_2 and O_2 seems straight-forward, but it also seems at odds with the notion that it can heat the terrestrial surface at the same time by radiation.

If CO_2 happily re-radiates (I know some don't like that expression) the energy it absorbs, why is the surface not simply tossing it right back out when it receives it?

Sure, the minerals of the surface will briefly become hotter but then they will radiate more vigourously and a percentage of that radiation will be at frequencies incompatible with CO_2 absorption, so it will be gone, gone.

What does "back-radiation" have to do with anything? Isn't it like radiation within an iron bar that I heat with an oxy-acetylene flame? Overwhelmed?

@marky48, including your IQ as part of your screen-name puts you at a bit of a disadvantage.

Very few will take your incredibly astute comments seriously and the planet will slip a little further into catastrophe. Yes, it's somewhat your fault but don't beat yourself up too much!

Nasif Nahle: @All...
Imagine you have 450 boxes on ground and each box can contain only one cat. Imagine you have only one box buoying in the air and it only can contain one cat. Now, imagine the Sun is sending 900 cats towards the ground, i.e. toward where the 450 boxes are placed. Remember, only one cat per box. From the 900 cats, only 450 cats hit on the ground, where the boxes are placed. 45 cats are "bounced" towards the outer space and 405 cats go into their respective boxes. There will be 45 empty boxes. But, ~365 cats

235

escape from their correspondent boxes and want to go back to the source of cats. They find that one box is obstructing their passage to the glory and one cat is "trapped", momentarily, in the box floating in the air.

Question: How many cats can be "absorbed" by the empty box that is floating in the air?

Let's continue: the cat inside the floating box tries to escape to the glory and jumps from the box.

Question: would the cat go back to the ground looking for another empty box? For answering this question, remember that there are not empty boxes on the ground because the Sun is sending cats continuously; therefore, the empty boxes on ground are occupied immediately by another cat incoming from the Sun.

Let's continue: the cat finds another empty box floating few meters far away from the first box from which it had escaped and tries to jump toward it. It fails and feels a strong cats' current from the ground pushes it towards the outer space; additionally, it finds that there is a very strong pressure exerted from the surface against which it cannot struggle.

Questions: will the cat be re-absorbed by any of the occupied boxes on the ground?

Could it be that other kinds of containers do exist in the air, at shorter distances than the prior box where it was "trapped"?

Could the cat prevail over that strong current of cats and the powerful pressure, which push it towards the outer space, if it doesn't bump into another empty container?

Keys:

Cats = IR quantum/waves.

Cats from the Sun=short wavelength IR quantum/waves.

Cats from the ground=long wavelength IR quantum/waves.

Cats jumping from the boxes floating in the air=very low energy density long wavelength IR quantum/waves.

Boxes on ground=limited and limiting configurations the energy absorbed by the ground can adopt.

Boxes floating in the air=limited and limiting configurations the energy absorbed by CO_2 molecules can adopt.

Other empty containers floating in the air=limited and limiting configurations that the energy can adopt without minimizing the free energy.

Cats' current=short and long wavelength IR quantum/waves outgoing from the emitters.

Pressure=radiation pressure.

Conclusion: Through spontaneous processes of heat transfer, it is impossible for a colder system to do work on another warmer system. In other words, it is impossible for a colder system to transfer energy towards a warmer system. :|

Smoking Frog: The Inclusion-Exclusion Principle requires that we know the overlap(s) to begin with, so it can't give us the answer.

> But it does give us an answer that is almost the same:
> .49 + .06 + .20 − (.49)*(.06) − (.06)*(.20) − (.49)*(.20) + (.49)*(.06)*(.20) = 0.61648.
> —Jim Masterson

It can't do it by itself, and you're not using it by itself. You're assuming random overlap without knowing it. (It didn't occur to me that a random-overlap calculation could be that simple, so I thank you for the insight.)

Let the cloud layers be called A, B, C, corresponding to the 49%, 6% and 20%. Assume that the patches that make up each layer are infinitesimally small and randomly distributed in the spherical shell in which the layer resides. Then the most likely (A,B) overlap is (0.49)(0.06), the most likely (A,C) overlap is (0.49)(0.20), and so forth for (B,C) and (A,B,C).

Naturally, the patches are not infinitesimally small, but I doubt that this changes the result by much. Non-random distribution could change it by much, but it would take a meteorologist to deal with that problem.

Anyway, your answer is better than mine. I can see that, generally speaking, the two methods will give different answers, but I don't see how to explain that the random-overlap answer is not the average of the two extremes.

Matt: What's up with all the boxes of cats?

The simplest way I know to understand the greenhouse effect is to realize that when you're standing on the ground in daytime the Sun is shining on you in visible light, and the atmosphere is shining on you in infrared. If you change the atmosphere so it shines more (e.g. adding more CO_2 and other greenhouse gases) you will be warmer on the surface because there is more electromagnetic radiation coming down on you than otherwise. That's it. More CO_2=more radiation on you.

All the complex logic might be important if you want to understand the details of radiative transfer (e.g. if you were grad student in physical oceanography or atmospheric science), but the previous description is good enough.

> I just say there is no scientific proof that increased CO_2 emission causes climate change, or that it does not. And I am not on any side in the debate, for that very reason.
> —Mikael Cronholm

Think for a bit. Is there scientific *proof* that Miami's climate next winter is going to be warmer than Minneapolis's?

Well, sure, because you can rely on basic physics. Because of the inclination of Earth's axis, in the winter there is more electromagnetic radiation hitting Miami than Minneapolis and this will be the case next winter too. You can handwave about advection and chaos and weather and unpredictability and yadda blah blah blah, but the overall number one physics controlling the temperature is the amount of EM radiation. After all, what else is there to make you above absolute zero? (only a tiny amount from radioactive decay of Uranium in the Earth's center, and this can be measured and it is miniscule).

More E&M radiation = warmer climate.

And likewise, more CO_2 in the atmosphere = more E&M radiation. This is not a theoretical prediction, this is a measured and observed *fact*.

With more CO_2 in the atmosphere it is physically impossible for the climate *not* to change.

Next step is quantifying the amount, and then it gets complicated, but the people who have been doing this for 50 years are reasonably good at it, and you should believe them.

Robert Clemenzi: George E. Smith says:

> *So whatever your particles are emitting downwards, they must be emitting a like amount upwards; which escapes to space.*

Actually, the lower atmosphere is mostly IR opaque with a few windows. For most of the spectra were absorption occurs, 99% of the available energy is absorbed within 20 meters of the surface. In the "wings" it requires about 500 meters to absorb 99%. From this, I deduce that very little radiation from the lower troposphere ever gets to space. Based on my analysis of lapse rate plots[140], it is pretty obvious that, near the surface, the atmosphere is IR opaque and the net radiation is toward the surface. Since it is opaque, the radiation emitted toward space is simply reabsorbed within a few meters. At the tropopause, the atmosphere becomes IR transparent and a significant amount of energy is emitted into deep space.

> *For the Sun to essentially emit a 5800K BB curve from the photosphere, one has many heavier elements present that are ionized and so are capable of emitting a continuum rather than merely a spectrum.*
> —*cba*

---

[140] my analysis of lapse rate plots

Actually, the line spectrum is temperature and pressure broadened to produce the continuous spectrum. Heavier elements are not required.

Mikael Cronholm: @ Matt. That does not sound right to me, as someone who deals mainly with terrestrial stuff. There is the Sun, there is a very long bit of near vacuum, there is the atmosphere, there is the Earth, and radiation goes from left to right in that sentence. Assuming the radiation from the Sun is constant, the near vacuum will be near perfectly transmitting, (leaving the atmosphere out for a while), we come to the Earth that will not transmit at all, so it can only reflect and absorb (and re-emit).

Now, are you saying that the transmissivity of the atmosphere will *increase* when there is more CO_2 in it? You will have to explain to me how, then!

The way I understand the theory is that the CO_2 supposedly lets the shorter wavelength solar radiation through, while blocking the longer wavelengths the Earth emits from radiating back into space, and that any possible decrease in the influx is counteracted by an even greater decrease in the outflux. And so the balance is disrupted, according to the theory. Or have I misunderstood something?

When you say "you should believe them" you begin to sound a bit like a missionary, and that makes me suspicious. Before I believe something I need to understand it.

And, another thing: "Think for a bit. Is there scientific *proof* that Miami's climate next winter is going to be warmer than Minneapolis's?" Well, yes, I think there is, and the point there is that the proof for the theory is repeated cyclically every year. The cycles that are discussed in the climate debate are much longer, if I am not mistaken, which means that there are no chances of repeating the "CO_2 test" and see if the same result occurs again.

Visible light will cause a temperature rise, hence heat an object, as it is absorbed. Google "Herschel experiment" and I am

Ken Coffman

sure you will find food for thought!

Another proof for that is that objects with different colors will absorb differently in the visual, white the least, black the most, and other colors in between. In the IR, the absorptivity (or emissivity) will NOT depend on color. And this is definitely my area of expertise, I assure you!

Bill Illis: On the blackbody ranking of 0 to 1 (with 1 being a perfect blackbody), what are the rankings for Nitrogen and Oxygen in the IR spectrum that Earth produces?

If they are not zero in the IR spec, what does that say about the radiation theory (considering that there is 35 times more N_2 and O_2 in the atmosphere than H_2O and CO_2).

Jim D and the radiation theory seems to assume they are Zero.

Matter: jae said:

> But, I would like to posit another "world" for the warmistas. Suppose the atmosphere of Earth consisted of ONLY N_2 and O_2. Now, these gases could not cool by IR emissions, but they would warm by conduction from the surface. And they could not radiate to space. So would they continually warm for millions of years?? Would they melt the planet? WTF, folks?

Heat would be transferred to the atmosphere by conduction and convection, increasing the energy in non-radiative transfers (perhaps it is kinetic energy or there is no change in angular momentum of the energy states so that the transition probability integrates to zero).

The atmosphere would warm, but as soon as it became as warm as the surface, it would stop absorbing heat (2nd law of thermodynamics—atmosphere can't get hotter than the surface!). The near surface atmosphere would approach the surface temperature, and the upper atmosphere would be cooler as defined by the dry adiabatic lapse rate (easily derived from PV = nRT plus

241

hydrostatic balance).

So no, it wouldn't warm forever. At equilibrium, it would have no net effect on the heat flows so the surface temperature would be the simple surface radiative balance (and much cooler than today, assuming ice is still allowed)

So the effect of conduction and convection eventually heads to an equilibrium value. Adding objects with an emissivity means that they absorb energy from the surface but they can also 'bypass' this absorption by increasing the emissivity higher up. However, since the surface is warmer than the higher level, it emits more. For an atmosphere where emissivity is not higher at higher altitudes then the effect is always warming.

We live in this world, hence the greenhouse effect warming us up.

Blade: Matt says:

> *The simplest way I know to understand the greenhouse effect is to realize that when you're standing on the ground in daytime the Sun is shining on you in visible light, and the atmosphere is shining on you in infrared. If you change the atmosphere so it shines more (e.g. adding more CO_2 and other greenhouse gases) you will be **warmer on the surface because there is more electromagnetic radiation coming down** on you than otherwise. That's it. More CO_2=more radiation on you.*

So the Law of Conservation of Energy has been repealed? Therefore Over Unity and Perpetual Motion devices are not only possible but practical? Electric cars that run off the alternator cannot be far behind. As another commenter pointed out, a vacuum would allow maximum transfer of EMR from the Sun. Any amount of matter in any form can only decrease the total transfer.

Now Matt, you nailed yourself to the wall right there for all the world to see. Seriously, right this minute you *should* be dwelling on the fact that the foundation you built all your conclusions on (and perhaps your religion) is nonexistent. An epiphany lies in your immediate future **if you are a wise person**.

However if you are otherwise, well, you will probably continue your journey on the AGW bandwagon. If so, I can recommend exciting offers that I find in my spam filter concerning ethanol, miracle fuel pellets and super duper windmills.

J. Bob: Eadler says:

> Calibration error doesn't affect the temperature anomaly if the error remains constant. In the case of equipment changes, so that the temperature trend becomes discontinuous, the change if detected is corrected for by the use of adjacent location data that is consistent. These adjustments are normally done by using computer programs. Calibration of the equipment is not really necessary.

First I'll agree this is somewhat off the thread, but temperature sensors include IR, and their results from the bases of these data sets, so much discussed. And much of what is discussed here relates to temperature and how accurate it is measured.

However if the sensors are in error, and that error is not accounted for, the resultant anomaly is subject to error. The reason for the Metrology Dept., is to correct for that slow insidious "drift" that takes place in instrumentation, which is why most reputable manufactures recommend periodic calibration. An earlier, and very well written item is:

http://wattsupwiththat.com/2011/01/22/the-metrology-of-thermometers/

As far as using adjacent stations to correct for others, it may be the "blind leading the blind", as noted by the quality of stations presented by the following site.

http://www.surfacestations.org/

As far as "adjustments" using computer programs, the old

saying "garbage in, garbage out" comes to mind".

barn E. rubble: Sensor operator says:

> *Do people really think climate change scientists want AGW to be true?*

Yes. Full stop. As stated previously, many of the best-known climate scientists have said as much. How else to justify your life's work? Why would M. Mann not switch to using mussel shells and such for proxies, instead of continuing with dodgy tree ring reconstructions? Because (I think) he's not interested in actually doing an accurate historical temperature reconstruction but instead wishes to find proof of something he really, really wants to find exists. For most warmongers I fear, AGW theory HAS to be true— too much invested to now say, "maybe not" in spite of what observation tells them.

> *If they are right, the end result is really bad.*

And that statement is based on what? Historically speaking. how much better things were when it was colder than when it got warmer? And please spare me the horrors of some areas (now considered a shit-hole that can't sustain life) COULD be worse.

If I hold my last utility bill in one hand and my wallet in the other, they will hit the floor at the same time if I drop them at the same time. Well established physics with centuries of consensus. Only what happens…?

Mikael Cronholm: IR is electromagnetic radiation. Or photons. The radiation is emitted when excited electrons shift from one shell to another, roughly speaking. The hotter an object is, the more excited the molecules and atoms are, and the more radiation will be emitted. When the radiation hits another object, the reverse happens. If that object happens to be you, you will feel the absorbed radiation as heat on the surface of your skin, and it will spread from there by conduction and convection (blood flow).

If you roast a chicken, does it heat up evenly all the way through at once? Of course not! It will absorb the radiation on the surface and then it will conduct (no convection, no blood flow, dead chicken, he-he) throughout the meat. A microwave oven heats from the "inside" though, because it excites water molecules directly, as they are dipoles.

Visible light will definitely heat things. So will parts of UV and all of IR. I don't just know it, I could prove it live to you, but not on a blog. So that ends my discussing of that issue.

Around the type of IR I work with, roughly 2-14 micron with the exception of the atmospheric absorption band at 6-8μm, human skin will absorb 98%. The rest is reflected. But my knowledge stops there, so maybe at even longer wavelengths there will be penetration, I don't know. In shorter wavelengths I would expect penetration to increase.

A penetration depth of UV of 1mm is something I would consider deep penetration in skin. At shorter wavelengths still, X-ray and gamma, we don't just get penetration, we have complete transmission through, except the very small portion that is absorbed.

If you ask Max Planck he will tell you that the energy of a photon increases with shorter wavelengths.

I don't know what kind of sauna you use, but it sounds more like a toaster. My sauna is a wet sauna, with a wood fired stove with rocks on it where I throw water to get steam. The heat transfer to my body is actually mostly through condensation, since my body is the coolest thing in there. So the latent heat from the steam is given off to my body as it condenses and mixes with sweat. There is radiation and convection taking place in there too, of course. But the real heat shock comes when I throw a good splash of water on the rocks!

Common sense does not always make scientific sense.

izens: Heat or energy in the form of photons or radiation travels in ALL directions including from cold to hot, otherwise it would be impossible to see yourself in a mirror which is colder than you are.

It is just that MORE heat travels from the warm region to the cold than travels from the cool to the warm so the NET flow of heat is from warm to cold. But the amount flowing back from cold to warm modifies the NET flow.

NoIdea: A thought experiment.

Let us construct a vessel in order to produce a simulated blackbody, this will essentially be a spherical chamber with an opening and a black interior. We shall construct it so that produces 1000W of continuous blackbody spectrum IR from the opening when at its desired target temperature.

Let us place it somewhere very cold, perhaps 2.7 Kelvin degrees and airless. (In orbit on the dark side of a dark moon perhaps?)

If we now heat our chamber to 193.18456 Kelvin, the IR leaving the opening should have its peak emission with a (continuous blackbody spectrum) wavelength at 15µm.

We are used to dealing with CO_2 in its gaseous form, from Wiki I found:

> At 1 atmosphere the gas deposits directly to a solid at temperatures below -78°C (-108°F; 195.1 K) and the solid sublimes directly to a gas above -78°C. In its solid state, carbon dioxide is commonly called dry ice.

I have not found a reference to the temperature that CO_2 deposits to at zero atmospheres, (or close to) I did find reference to the fact that dry ice has been detected on Mars and in comets. So for now I will assume that it can exist in a near vacuum.

If we now introduce a chunk of CO_2 (say 1kilo at 3 Kelvin) in front of the opening to absorb and get heated by this 15 µm peak IR, how warm could it get?

Nasif Nahle: @Matt...What's up with all the boxes of cats?

The simplest way I know to understand the greenhouse effect is to realize that when you're standing on the ground in daytime the Sun is shining on you in visible light, and the atmosphere is shining on you in infrared. If you change the atmosphere so it shines more (e.g. adding more CO_2 and other greenhouse gases) you will be warmer on the surface because there is more electromagnetic radiation coming down on you than otherwise. That's it. More CO_2=more radiation on you.

The cats are an illustration about what actually happens through the heat transfer from the Sun to the ground and from the ground to the air, as well as the impossibility that a colder system warms up to a warmer system. A scientist must adhere to observed phenomena and to well-tested theories.

The simplest way you know is not the correct mechanism how it actually happens, though popular; however, popularity is not a step of the scientific method.

The atmosphere doesn't shine. The Sun does shine.

Matter: @Blade, there is no breaking of conservation of energy; to everyone who doesn't believe in the greenhouse effect you have a lot more reading to do: I recommend Guenault's statistical physics textbook as an excellent (and concise!) starter. Atmospheric Physics by Ambaum is also concise and deals with radiative transfer in later chapters. Most of your claims are dealt with explicitly there, if you can follow the maths.

The emission ability of a substance is related to its absorption (if it has energy levels allowing emission, it has the same levels allowing absorption). This value depends on wavelength for most materials — you can be transparent to visible light but absorb a lot in the infrared for example.

The emissivity/absorptivity of the atmosphere in the wavelengths sent by the Sun is relatively small (with notable exceptions like ozone absorbing UV). CO_2/H_2O doesn't absorb here so adding it doesn't reduce the amount of energy hitting Earth.

But it does absorb in the lower wavelengths. Since it absorbs there, it must be able to emit there too. The atoms move quickly and interact with each other, which allows CO_2 to dump heat into the O_2/N_2 and vice versa.

Adding CO_2 doesn't prevent sunlight hitting Earth, but it does absorb light going up from Earth. Hence the warming.

Will: Mikael, did you see Bill Illis' excellent question, and do you know the answer?

> Around the type of IR I work with, roughly 2-14 micron with the exception of the atmospheric absorption band at 6-8um...

Is 6-8μm the absorption band of O_2 and/or N_2? Because it isn't CO_2 is it? CO_2 absorption is 15μm as we all know.

So when you refer to "the atmospheric absorption band at 6-8μm" which part of the atmosphere is absorbing at 6-8μm please?

A C Osborn:

> In fact the warmer the surface becomes the more the energy will be radiated to space without bouncing around in the atmosphere.
> WUWT?
> So the surface has a direct connection to space then?
> Oliver Ramsay
> Yes

Wow, I didn't realize there were any Mountains that high. The Earth's Surface is in DIRECT contact with space?

Not going through any atmosphere for the photons to "bounce around" in then?

kuhnkat: A C Osborn says:

> *Wow, I didn't realize there were any Mountains that high.*
> *The Earth's Surface is in DIRECT contact with space?*
> *Not going through any atmosphere for the photons to "bounce*
> *around" in then?*

AC, you should try a job as a comedienne, or not. You apparently need to read about atmospheric windows.[141]

I'll leave it to others to decide whether particles can reach states in the atmosphere where they can no longer absorb in bands where they typical are considered to absorb.

Domenic: To Will: Water. H_2O absorbs nearly 100% in the 6 to 8 micron band. If you are an IR device operator, and you wish to estimate temperature measurement with any degree of accuracy from more than a few inches away, you had better not be using a filter that includes the 6-8 micron band.

CO_2 is practically a non-issue in IR thermography (using IR cameras). CO_2 is irrelevant.

That is one of the points that I keep making. Most physicists, scientists, who keep trying to point to CO_2 as a major contributor to the greenhouse effect, have never spent time in the REAL WORLD working directly with IR and trying to quantify measurements with it. If they did, they would quickly see the foolishness of their assumptions regarding CO_2, and would be amazed at the magnitude that H_2O truly plays.

Forget CO_2. It's so minor that it is nearly irrelevant.

Michael Moon: After wrestling with this question of CO_2 heating the atmosphere, my take is this: CO_2 absorbs at several different wavelengths. 15 microns is the one where water vapor isn't already absorbing all of Outgoing Longwave Radiation. 15 microns

---

[141] http://csep10.phys.utk.edu/astr162/lect/light/windows.html

wavelength is in the far infrared, corresponding to -4 degrees F. So, when the ground is at -4°F, COLDER CO_2 in the vicinity can absorb radiation and be heated. If the atmosphere is warmer than the ground, all bets are off.

Now, since the atmosphere radiates too, any CO_2 ABOVE air at -4 F, no matter how high up over the ground, can also absorb radiation and be heated. Of course, when CO_2 absorbs 15 micron radiation, the molecule of CO_2 is excited, and the atoms vibrate. Due to Brownian motion, this vibration is immediately absorbed as extra kinetic energy (=heat) by surrounding molecules. So, if the ground is at -4 F, the 4-parts-in-10,000 CO_2 can absorb and transfer a little heat to the atmosphere without violating the Second Law.

It doesn't seem enough to get all excited about. Water can freeze at 59°F ambient outdoors with quiescent atmosphere and a very clear night—it is in my Thermo textbook.

George E. Smith: Matter says @Blade:

> The emission ability of a substance is related to its absorption (if it has energy levels allowing emission, it has the same levels allowing absorption). This value depends on wavelength for most materials—you can be transparent to visible light but absorb a lot in the infrared for example.
>
> The emissivity/absorptivity of the atmosphere in the wavelengths sent by the Sun is relatively small (with notable exceptions like ozone absorbing UV). CO_2/H_2O doesn't absorb here so adding it doesn't reduce the amount of energy hitting Earth.
>
> But it does absorb in the lower wavelengths. Since it absorbs there, it must be able to emit there too. The atoms move quickly and interact with each other, which allows CO_2 to dump heat into the O_2/N_2 and vice versa.
>
> Adding CO_2 doesn't prevent sunlight hitting Earth, but it does absorb light going up from Earth. Hence the warming.

Ken Coffman

Well I can't tell what @Blade said from what Matter said; but there's enough misinformation to go around here.

> The emissivity/absorptivity of the atmosphere in the wavelengths sent by the Sun is relatively small (with notable exceptions like ozone absorbing UV). CO_2/H_2O doesn't absorb here so adding it doesn't reduce the amount of energy hitting Earth.

Take that for example; demonstrably false; and not just a little bit false. For a start, the solar spectrum received at Earth contains 98% of its energy essentially in the wavelength range from 0.25 microns in the UV to 4.0 microns in the IR, with only 1% left over at each end. Ozone of course cuts off the short end at about 0.3 microns. About 2.5% of solar radiation lies below 300nm. Ozone also takes a small bite out of sunlight at around 0.6 microns; but not much. CO_2 on the other hand has several significant absorption bands in the solar spectrum IR region; first at 2.0 microns; where normal atmospheric CO_2 absorbs about 35% of that band, then there is a strong 2.7 micron CO_2 band where essentially 100 % is absorbed, and finally there is an equally strong CO_2 band at 4.0 microns. The main GHG CO_2 band is of course in the region of 13.5 to 16.5 microns.

For the 4.0 micron CO_2 band; the solar spectrum has only 1% left in that area, so a fairly small but NOT zero CO_2 absorption of incoming solar energy. At 2.7 microns, the solar spectrum still has 3% of its energy left; or 41 W/m^2, so CO_2 can capture a good part of that (which therefore will NOT reach the ground as incoming solar energy.

Then there is that pesky non GHG feedback H_2O. It has a strong absorption band starting at about 4.0 microns, and going up to about 6.5 microns, where the "Atmospheric Window" starts, so it overlaps the long edge of the CO_2 4 micron band so doesn't have much impact on that 1% remaining solar energy. BUT H_2O has a strong band that starts at about 2.3 microns, and goes to about 3.3 microns completely enveloping the CO_2 2.7 micron band. (but note that the high resolution lines are likely to be separated; but a good chunk of that solar energy will fall to water or CO_2.

But water really gets going at shorter than the 2.3 to 3.3 micron band. There are bands at 1.6-1.8, 1.2-1.4, 0.94, 0.85, and 0.75 microns.

47% of the solar spectrum energy lies at longer than 0.75 microns wavelengths, and water plus CO_2 could easily account for about half of that or say 20% of the total solar incoming energy can be absorbed by H_2O and CO_2 in the atmosphere, and not ever reach the surface as solar spectrum energy, where much of it would be deposited deep in the oceans.

So it simply is NOT true, to say that "CO_2/H_2O doesn't absorb here so adding it doesn't reduce the amount of energy hitting Earth.

Every single additional molecule of H_2O or CO_2 that is added to the atmosphere WILL reduce the amount of incoming solar energy that reaches the surface of the planet to warm it.

Yes that energy will warm the atmosphere as a result; and that warmed atmosphere; being above absolute zero, will radiate a thermal LWIR spectrum, and only half of that will be directed down towards the ground, so the rest will get lost to space.

Thermal radiation depends on temperature; not on energy levels either atomic or molecular; it is a result of the thermal motion (and acceleration) of electric charge. Just think about the acceleration of charge that occurs when a warming GHG molecule collides with an ordinary gas molecules or atom (N_2, O_2, Ar). So every collision that results from the non zero temperature of the gases, results in charge acceleration, and consequent Electromagnetic Radiation according to the Planck and Stefan-Boltzmann equations; not that you necessarily get a complete black body spectrum; but what you DO get, has that Planck BB spectrum as its outer envelope (can't exceed).

People keep coming here to WUWT and insisting that gases

do NOT emit thermal radiation. They are about the only people who believe that. Physicists don't believe that; because they have never not observed such non-emission. Hot things radiate thermal spectra; they don't suddenly stop doing so when they melt and vaporize, or simply sublime.

"Heat" is NOT energy in the form of photons. Heat is purely MECHANICAL KINETIC ENERGY in the form of molecular or atomic vibrations due to collisions between particles (of matter) Sans matter; there is NO "HEAT". Heat is characterized by TEMPERATURE.

Photon energy is Electromagnetic Radiation that obeys Maxwell's equations; from down to; but not including DC; and extends all the way beyond the gamma ray spectrum.

EM radiation and photons don't know ANYTHING about Temperature; and they can go where they damn well please; hot or cold.

The very same Earthshine radiation that leaves Earth for the dark side of the moon will if it misses the moon proceed on and could reach the Sun which has the same angular size as the moon, so the amount of Earthshine that could hit the moon can also hit the Sun.

Leif Svalgaard, has not reported ANY instances, of Earthshine photons being refused landing permission on the Sun; despite its 6,000 K temperature.

And the second law of Thermodynamics as stated by Clausius, refers to cyclic machines:

> *No cyclic machine can have no other effect, than to transport "HEAT" from a source at one Temperature, to a sink at a higher Temperature.*

...doesn't say anything about EM radiation; just "HEAT".

We wouldn't have this problem if people just accepted that "HEAT" is a verb, NOT a noun.

Oliver Ramsay: A C Osborn says:

> *Wow, I didn't realize there were any Mountains that high.*
> *The Earth's Surface is in DIRECT contact with space?*
> *Not going through any atmosphere for the photons to "bounce*
> *around" in then?*

Well, AC Osborn, it's too bad your humour is commensurate with your knowledge and not with your truculence. Since kuhnkat has given you a pointer to the physics, I'll address your language usage and comprehension. You asked about "direct connection" and then got it all confused with "contact".

I know it's difficult because words have literal and figurative and contextual meanings, but with patience and perseverance you will get the hang of it.

ThomasU: I had asked earlier if it was possible to actually measure the energy balance of our planet. P. Wilson—thanks for the reply—pointed to the fact, that temperature measurements are no good for this purpose.

What I had in mind was a satellite which should be able to actually measure the radiation coming to Earth from space and also measure the radiation going from Earth to space. Is it at all possible to do this? Has it already been done? Spectral analyses of this kind should be possible I guess, I'd almost expect they were already made. If so: is there anybody around who can comment the results? It seems to me that a great deal of the theories which are claimed to substantiate the AGW hypothesis suffer from a frightful lack of evidence. I see the so called "glass-house effect" as a radiation delay. The gases in the atmosphere delay the radiation to space. If there was no delay, nighttime cooling would be similar to that on the moon, I guess. The main source of energy on Earth is the Sun. If its output gets less—for whatever reason—the Earth is going to receive less and get cooler. And vice versa.

Well this discussion being as long as it is, I could say a lot

Ken Coffman

more, but I want to come back to my question: Has the radiation (or energy) balance of Earth ever been measured from space? Is it at all possible?

Matter: George E Smith, I made a humiliating series of errors there and apologize for my last post, it should be deleted.

My CO_2/H_2O energy approximation was a mistake from a quick misreading of the waveband units, absolutely stupid of me.

My thermal radiation mistake was from assuming a constant temperature and looking at changes in emission, which filters out the non-emissivity related features so I associated them with internal level energy transfers.

I took my conclusion from the results of line-by-line radiative transfer models along with measurements of net heat flow as a function of wavelength, which shows that a greenhouse effect does exist and that CO_2 strengthens it. My understanding of the in-between bits was wrong. Thanks for pointing me in the right direction George; I'll take a bit more time out to check through the full spectral results now. :)

barn E. rubble: Domenic says:

> That is one of the points that I keep making. Most physicists, scientists, who keep trying to point to CO_2 as a major contributor to the greenhouse effect, have never spent time in the REAL WORLD working directly with IR and trying to quantify measurements with it.

That statement I find most interesting. Are there any arguments to the contrary? If I understand correctly, the debate or questions about feedbacks (+/-, &/or whatever) caused by increased CO_2 levels (manmade or otherwise) isn't really moot but trivial?

Konrad: ThomasU says: The short answer is that it is possible to measure the in going and outgoing radiation from Earth. A satellite was built to do this named Triana later renamed DSCOVR. The satellite would have orbited Earth in the L1 position, giving it a whole hemisphere view. There has been much debate about why this satellite was not launched with both sides of U.S. politics claiming that the other side were afraid of the results. I believe the satellite, which was delayed by problems with the space shuttle, has been approved for reconditioning by the present administration for launch "sometime" in the future.[142]

At present the work by Dr. Spencer utilizing existing lower orbiting satellites, has indicated that there is little radiation imbalance, and that any water vapour feed back to increases in temperature are negative. The DSCOVR satellite would have however provided far better data to analyze.

Will: Sorry, Domenic no offence, but my question is to Mikael, not you. You have missed the point with your H_2O response. There is nothing in Mikael's statement that indicates he is referring to water. He states clearly "atmospheric absorption band at 6-8μm".

This is clearly not a reference to H_2O.

The atmosphere does not consist of H_2O which, incidentally, absorbs strongly in many regions not just the 6-8μm.

So when you butt in and answer questions that are not directed at you, you miss the point and ultimately muddy the water. Whether this is deliberate or simple bungling interference, the end result is the same. You have diluted a pertinent question with irrelevant distractions.

So I ask once more, and this question is to Mikael (and not anyone else!) Did you see Bill Illis' excellent question, and do you know the answer?

Quote: "Around the type of IR I work with, roughly 2-14 micron

---

[142] http://en.wikipedia.org/wiki/Deep_Space_Climate_Observatory

with the exception of the atmospheric absorption band at 6-8μm"
Is 6-8μm the absorption band of O_2 and/or N_2? Because it isn't CO_2
is it? CO_2 absorption is 15μm as we all know.

So when you refer to "the atmospheric absorption band at 6-8μm" which part of the atmosphere is absorbing at 6-8μm please?

Domenic: To barn E. rubble: Yes. CO_2 is trivial.

To Will: I noticed Mikael has a day job, and would not be able to answer your question for a while. So, I gave you the answer to keep the thread going. It is interesting to me to see what kind of understanding of IR physics exists in those concerned about AGW and the greenhouse effect.

Mikael Cronholm: @ Will. Your comment was not lost, but I was sleeping. I live in Thailand so I am on an almost opposite time zone compared to the U.S.

Anyhow, the absorption rates and wavelength bands of N_2 and O_2 are unknown to me; they are accounted for in the models we use together with all the rest in the air. Domenic correctly answered your other question about the absorption band that separates the two wavebands we use for thermal imaging, the dominant absorber there is water vapor.

I will explain a little deeper about the atmospheric compensation that is done when we measure temperature in IR. Two spectral bands are used, 2-5 and 8-14μm, approx. These days the longer one dominates the industry completely, for practical reasons and cost, because those detectors are un-cooled. (In furnaces we use a narrow band at 3.9μm to avoid H_2O and CO_2 generated by the flames from the burners. That is an extreme atmosphere.)

Remember, "atmospheric" does not mean we deal with higher altitudes. We assume sea level, I guess. The calculation uses the LOWTRAN atmospheric model, which is an empirically developed model. The camera has three inputs for the compensation; distance, air temperature and relative humidity. Air temperature is used for two things, to re-calculate the relative humidity to an absolute value, and to determine the emission from the atmosphere. The atmospheric emission, once calculated, is removed from the total signal. Then there is the absorption, the atmosphere gives and takes, so the calculated absorbed radiation is put back. Simple as that!

Notice that the model is empirical. That tells me that the scientists that figured all this out at some point and did not consider it worthwhile to try and calculate the influence of all the gases involved.

Notice also that although we are not in the H_2O absorption band, humidity—water vapor—is still the only gas that is individually accounted for in the model, because it is the only one that changes significantly from time to time. That means the other gases are just a standard soup with an assumed recipe. So that is where O_2 and N_2 are accounted for, they are included in the empirical LOWTRAN model and their individual contributions are unknown to me, because I don't need to know them to do my job.

Now, if I would be manipulating these numbers to see what errors are involved, the one that throws the measurement off the most is distance. So the compounded effect of all gases together seems as large as H_2O itself. Air temperature is the second most important, but seldom gives any large errors. Relative humidity is almost insignificant, it can usually be left at 50% and forgotten. I put it at 25% or 75% if I think I am closer to those values, but I never make a humidity measurement to be exact. It changes only on the decimals, if at all. This is valid in the distance ranges we normally work, up to about 20 m.

What we have to remember about absorption bands is that they do not abruptly cut on and off, they vary up and down with wavelength. I don't think there is any gas that is COMPLETELY transparent in ANY part of the IR spectrum at least. And as a GENERAL rule I always assume that the shorter the wavelength the better the penetration. The band variations that exist I think have to do with optical effect on a molecular level due to the geometry and composition of the molecules, but I am not an expert on that, so don't take my word for it. If anyone has initiated input I would appreciate it.

@ George E. Smith

I just want to say I appreciate your input in this comment a lot. I have that solar spectrum curve in front of me if I close my eyes. And when I remind myself that the darn thing is in a logarithmic scale, all those numbers you give make a lot of sense. Thanks for that!

My only slight objection is here:

> *"Heat" is NOT energy in the form of photons. Heat is purely MECHANICAL KINETIC ENERGY in the form of molecular or atomic vibrations due to collisions between particles (of matter) Sans matter; there is NO "HEAT". Heat is characterized by TEMPERATURE.*

Agreed to the most part. It is the last sentence I am a little bothered with. Heat and temperature are connected in such a way that they co-vary depending on heat capacity, BUT with the exception of latent heat during phase change. If you observe a temperature change you can safely assume that the amount of heat in the object has also changed, but if you have a change in heat it may not necessarily cause a change in temperature, if there is phase change going on.

Domenic: I must admit I am impressed with your patience in answering the many questions here. The misconceptions about IR and radiational heat transfer are so vast in the scientific community...

IR cannot penetrate into the body. It is absorbed by the skin to a minor depth. Most of skin composition is H_2O. That is the main reason why.

H_2O is a powerful absorber and emitter of long wavelengths.

When you look at an IR image of a person's head, you can see some arteries and other features. But it is not because the IR looks within the body, but rather because there are certain arteries that are close to the skin surface. And there are parts of the surface of the human body where there are lots of tiny arteries just below the skin, high blood perfusion, that show up as large hot spots.

This may help some begin to understand what radiational heat transfer really is. Take a look at this graphic:

http://en.wikipedia.org/wiki/File:EM_Spectrum_Properties _edit.svg

Now, I really like this graphic because it uses a nice handy old fashioned thermometer figure at the bottom. So you have a nice easy way to look at the idea.

If you have a perfect 'blackbody' atom, at any temperature over absolute zero it emits radiation (heat energy). That perfect blackbody atom near absolute zero will first emit very, very long radio waves. As you heat that blackbody atom up, the hotter it becomes, the heat energy shifts to shorter and shorter wavelengths.

Just follow the thermometer and the wavelengths. That graphic is perfect in illustrating what basically happens in radiation in an ideal situation.

Everything emits energy above absolute zero. Everything. In effect there is no such thing as COLD. There is only less heat and more heat. Heat flux, or radiational heat transfer, is simply the NET flow of heat (or energy) from a hotter material to a less hot material. The forms that 'heat radiation' are classified as, is exactly as shown in the graphic.

But, in effect, all matter in the universe is in a form of constant sort of 'communication' due to electromagnetic radiation, but with net energy moving from hotter to less hot.

However, real world atoms (hence molecules, etc) are not perfect blackbodies, so they will GENERALLY follow this graphic, subject to unique quirks of each different type of atom or molecule. Those quirks are the tricky part. And science has not tested various atoms and molecules anywhere near enough to fully understand those quirks. Even N_2 which makes up most of the atmosphere has hardly been tested at various temperatures, thus wavelengths to fully understand the nature of its true radiational properties.

Mikael Cronholm: IR is a very broad spectrum. Some IR may penetrate deeply, you may want to inform me exactly which wavelengths, if you know, I am truly interested (not sarcastic, I promise). But certainly NOT the kind of IR we look at with thermal imagers. From Hollywood, people "know" that we can see through walls and stuff with IR. That is probably where that BS comes from originally. It is BS though, no less.

If you don't believe condensing steam will give off huge amounts of latent heat, put the kettle on and when it boils, stick your finger right in the steam that comes from it and you will find out.

Most paints and dyes are approximately "grey" in IR images, if we refer to "black" as being blackbody absorbing. Paint will be about 0.90-0.95 emissivity in a long wave camera.

Glass is what is referred to as a selective radiator. It means that its emissivity, reflectivity, and transmissivity all change with wavelength. Visible light passes through it easily, but there is a change in the composition of the spectrum on the other side. There is. Not so much in visual, but in IR there are big things going on! We have to forgive Herschel, he had no spectrometers at hand, but in this day and age we do. Glass will start to lose transmissivity around 2µm and nosedive completely at 2.3µm.

Over 3µm there is hardly any transmissivity AT ALL!!!!!! How can I know for sure? Because glass is totally opaque to any IR camera I use, in addition to all the spectral charts that are available. Can't see through it at all. Had Herschel moved his thermometers a little further from NIR, into where IR should have been, they would have dropped completely. His result can only be seen as qualitative ("there is other light than visible"), rather than quantitative.

FYI, in IR cameras we use lenses made of either silicon (3-

5μm) or germanium (8-14μm). Glass lenses would definitely not work.

Knowing that no IR over 2.3μm will go through a normal window, we should accept that the heat you feel when you stand behind a window where the Sun shines in comes from visual light.

"Warm" or "cool" colors may exist in psychology, but not science. If we think red is "warmer" than blue for example, that concept would be backwards, since blue incandescent light would need a much hotter object to dominate the spectrum. An object at 7245K would peak at 0.4μm, for example—much hotter than the Sun. Consider arc welding and notice how blue the arc looks. Very hot, that's why!

@ George E. Smith. I noticed afterward that all those nice numbers were in a previous post, not the one I referred to. Very useful nonetheless. Thanks!

Nasif Nahle: George Smith says:

> Every single additional molecule of H_2O or CO_2 that is added to the atmosphere WILL reduce the amount of incoming solar energy that reaches the surface of the planet to warm it.

That assertions is not valid for CO_2, which is an absolutely-blind cat to short wavelength IR quantum/waves incoming from the Sun.

It's a one-eyed cat towards long wavelength IR quantum/waves; nevertheless, it only absorbs a little from this spectral band because its absorptivity, taking into account the mean free path length and the crossing lapse time of IR quantum/waves IN the atmosphere, is quite low: 0.004. Taking into account the whole column of air, the CO_2 Pp and actual temperatures of the atmosphere—not that of an idealized blackbody because CO_2 is not a blackbody neither it "resembles" a blackbody, the total emissivity of the carbon dioxide is no more than 0.002.

As a way of comparison, the total emissivity of water vapor is 0.7, the total emissivity of moistened clay with organic materials is

0.95. The total emissivity of liquid water is 0.96. These observed and verified facts transform the CO_2 into a feeble one-eyed cat.

Alternatively, the delay that a molecule of carbon dioxide causes to the emission of electrons and photons, after having absorbed a photon, is measured in attoseconds (as), i.e. 10^{-18} of a second; consequently, there is no way for the atmospheric carbon dioxide to "hold" the absorbed energy for periods longer than 20 ± 5 as. : |

Blade: Matter says:

> @Blade; There is no breaking of conservation of energy; to everyone who doesn't believe in the greenhouse effect you have a lot more reading to do: I recommend
> {... blah, blah ...}
> Adding CO_2 doesn't prevent sunlight hitting Earth, but it does absorb light going up from Earth. Hence the warming.

Very sloppy on your part, Mr. Matter. You immediately got sidetracked from the simple and iron-clad point I made (and as a commenter mentioned) you didn't bother to quote me which provided cover for your conversational detour. So, let's recap:

Matt says:

> The simplest way I know to understand the greenhouse effect is to realize that when you're standing on the ground in daytime the Sun is shining on you in visible light, and the atmosphere is shining on you in infrared. If you change the atmosphere so it shines more (e.g. adding more CO_2 and other greenhouse gases) you will be **warmer on the surface because there is more electromagnetic radiation coming down** on you than otherwise. That's it. More CO_2 = more radiation on you.

Then I said: "So the Law of Conservation of Energy has been

repealed? Therefore Over Unity and Perpetual Motion devices are not only possible but practical? Electric cars that run off the alternator cannot be far behind. As another commenter pointed out, a vacuum would allow maximum transfer of EMR from the Sun. Any amount of matter in any form can only decrease the total transfer."

I'll make it crystal clear...

Let the Earth have no atmosphere and we shall call the EMR striking the Earth as **X**,

Let the Earth have its present or any atmosphere and we shall call the EMR striking the Earth as **Y**,

I ask you now: Is ... **X < Y** ... or ... **X = Y** ... or ... **X > Y**? There is only one answer. **X > Y**.

P.S. Therefore your last sentence is wrong...

Adding CO_2 doesn't prevent **sunlight** *hitting Earth, but it does absorb* **light** *going up from Earth. Hence the warming.*

(you certainly employ vague yet enigmatic use of the terms sunlight, and then light wouldn't you say?)

Let's avoid confusion. If a single photon of **any part of the EMR spectrum** is either reflected or absorbed and re-radiated to space, (i.e., **it never gets here!**), then, the net EMR is reduced, period. The point (as I was responding to Matt) is that there is no way that any matter (Atmosphere) placed in between the Sun and the Earth can ever increase the amount of EMR we receive. It is theoretically possible (but highly unlikely) for it to be EQUAL, but not GREATER.

Now unless you are planning on telling us that CO_2 does not intercept a single photon of any part of the EMR spectrum traveling from the Sun to the Earth, (not a single photon? CO_2 is a perfect insulator?), then only one statement can be held as true: CO_2 decreases the net EMR transmitted from the Sun to the Earth (even if it is a single photon). You certainly are not proposing there is no downstream infrared from the Sun are you?

The current controversy seems to lie in the energy balance of the photons that actually make it here through the atmospheric 'wall'. On this I agree with the crude blanket analogy, the energy bounces around here a little longer than it would were there no atmosphere, CO_2 or not. That means that if you switch off the Sun, we'll take just a little longer to freeze to death. How much longer? Who cares? Not long enough though.

Richard Sharpe: George E Smith seems to be saying that CO_2 (and H_2O) are double edged swords.

That is, every extra molecule of CO_2 contributes to keeping some fraction of incoming radiation from making it to our surface (approximately half of what is intercepted is re-radiated, although some of it contributes to heating the atmosphere), while it also contributes to keeping some of the outgoing energy in the atmosphere for a while longer.

George E. Smith: Richard Sharpe says:

> *George E Smith seems to be saying that CO_2 (and H_2O) are double edged swords.*
>
> *That is, every extra molecule of CO_2 contributes to keeping some fraction of incoming radiation from making it to our surface (approximately half of what is intercepted is re-radiated, although some of it contributes to heating the atmosphere), while it also contributes to keeping some of the outgoing energy in the atmosphere for a while longer.*

Richard, that is precisely what I am saying BUT!! Let's keep a proper perspective. I think the "back of the envelope" chicken scratchings I did, show that perhaps CO_2 is not a big culprit as far as lowering the incoming; but the effect is incontrovertible; it is real. The issue is much more important for H_2O, because there is a lot more of it; it is never less than the amount of CO_2, and it varies widely geographically, and with weather.

In the long haul, there can be no doubt, that an increase in H_2O on a global scale, and over climatically meaningful time scales; must reduce to total solar energy that reaches the Earth surface; where it can be stored in the oceans, or in the land materials. Sooner or later, that must mean a cooler Earth.

I should add one further clarification. The main solar spectrum portion that is intercepted by H_2O, and the same is true for the CO_2, is longer than 0.75 microns, and as long as 4.0 microns. It would NOT be correct to say that this is deposited "Deep" in the oceans, like the bulk of the solar energy is.

As every scuba diver knows, the red end of the spectrum disappears first as you go deeper, and 750nm is a quite deep red; so we are talking metres to tens of metres; but not hundreds. We know that H_2O has its very strongest spectral absorption at 3.0 microns, where maybe a couple of percent of solar energy still is to be found. At that wavelength the water absorption coefficient, is about 9,000 cm^{-1}, so the 1/e absorption depth is just 1.1 microns, or 5.5 microns for 99% extinction.

So to some extent much of that lost solar energy, would have been absorbed in shallower water, and would likely have given rise to enhanced evaporation from the surface; which is pretty much what the returned LWIR from the atmosphere does also.

We should note that both GHG absorption of LWIR, and similar absorption of incoming solar energy HEAT THE ATMOSPHERE. They do NOT heat the ground. Once the atmosphere is heated; regardless of means; the ambient LWIR radiation which is going in all directions, is a function only of that atmospheric temperature; and quite independent of what heated the atmosphere.

As others constantly point out here; convection and evaporation and other energy transport mechanisms are perhaps more important to the Earth cooling process, at least at living elevations. I don't ignore those effects; I just don't dwell on them; since absolutely none of that is in any way affected by GHGs or the Greenhouse effect. That is purely a radiation phenomenon; which is why I concentrate my thoughts on the radiant energy processes. I'm happy that others do keep track of the other thermal processes, and the various circulations like ENSO and AMO etc.

Water really is THE double edged sword, since it is the only condensing GHG; and in liquid or solid form as clouds, I'm convinced it ALWAYS cools the Earth; NEVER warms it. Those high clouds that we associate with humid balmy nights; are a consequence of the earlier surface conditions (Temperature and Humidity); they are NOT the cause of those surface conditions.

Ken Coffman: I think a lot of my own confusion comes from buying into the warmist strategy of counting photons: up, down and sideways. You get yourself into a tangled mess and into the mysterious "back radiation" and equally mysterious "net radiation". We have a choice—we can think of photons as particles or waves. So, what if we imagine waves instead of photons? When we do this, I think the picture becomes much more clear.

> The amount of radiation emitted from each of them depends
> on two things ONLY, the temperature of the object and its
> emissivity. So radiation is not a side effect to temperature, it is
> THE EFFECT.
> —Mikael Cronholm

If we think of waves like lines of flux, then they point from the warm body toward the cold body (and there are none when the bodies have equal temperatures). We have radiation and nothing like anti-radiation. Given this mental model, (ignoring temperature inversions for now) is there any way for atmospheric CO_2 to increase the temperature of the Earth's surface? It can modify the cooling rate, but it cannot increase the temperature. So, global warming with more and more record high temperatures? This is not something CO_2 can do.

This is a point made over and over from different points of

view in the *Slaying the Sky Dragon*[143] book. I, for one, am looking forward to Mikael's comments about this book, good, bad or indifferent.

George E. Smith: Mikael Cronholm says @ George E. Smith:

> *I just want to say I appreciate your input in this comment a lot. I have that solar spectrum curve in front of me if I close my eyes. And when I remind myself that the darn thing is in a logarithmic scale, all those numbers you give make a lot of sense. Thanks for that!*
>
> *My only slight objection is here:*
>
> *"Heat" is NOT energy in the form of photons. Heat is purely MECHANICAL KINETIC ENERGY in the form of molecular or atomic vibrations due to collisions between particles (of matter) Sans matter; there is NO "HEAT". Heat is characterized by TEMPERATURE.*
>
> *Agreed to the most part. It is the last sentence I am a little bothered with. Heat and temperature are connected in such a way that they co-vary depending on heat capacity, BUT with the exception of latent heat during phase change. If you observe a temperature change you can safely assume that the amount of heat in the object has also changed, but if you have a change in heat it may not necessarily cause a change in temperature, if there is phase change going on.*

Mikael, what I meant by that statement, is simply this. "HEAT" (energy) is nothing more nor less, than the total kinetic energy of all of the particles of matter in the "sample". That total energy, and its distribution as to particle velocity is defined by the temperature. The Maxwell-Boltzmann distribution of velocities is defined by the temperature (in gases).

---

[143] *Slaying the Sky Dragon: Death of the Green House Gas Theory*, published by Stairway Press, 2011

Ken Coffman

The average KE per particle is simply 3kT/2. That is for the translational energy (in three axes). Some molecules will also have rotational degrees of freedom, so they can have more energy per particle, and one can calculate the RMS velocity as sqrt(3kT/m) where m is the particle mass.

One can calculate the velocity distribution (ideal gas) in the form:
$-(1/N)dN/dv = 4\pi v^2(m/2\pi kT)^{3/2} \cdot \exp(-mv^2/2kT)$ which is the M-B distribution.

As to these curves, my most useful source of (somewhat dated) thermal data is *The Infra-Red Handbook*, and I have several Optics Handbooks that I use at work all the time, that have a lot of stuff. Another VERY useful curve to have is a Normalized Back-Body Radiation curve. Normalized in the sense that the wavelength axis is normalized to the peak wavelength, so the peak is at 1.0, and the spectral radiant emittance is also normalized to the value at the peak wavelength so it also is 1.0.

There is such a curve in *Modern Optical Engineering* by Warren J. Smith; published by McGraw Hill. It is his Fig 8.7 I have an old copy and also a new copy and the curve is in both. He plots wavelength from 0.1 to 50 (times the peak) and relative spectral emittance from 0-1 (linear) and also from $1.0\text{-}10^{-5}$ (logarithmic), and then he has a nifty scale of fraction of total energy emitted below, going from 10^{-6} (at 0.24 of peak wavelength) up to 99% at 8.0 times peak.

So I don't really have to eyeball that much; I can read it off the graph. Since we can't cut and paste pictures here, it is hard to show people this data. But if you care to give Anthony permission to send me your e-mail address, I'd be happy to snap some digital photos of some of these useful graphs, and e-mail them to you.

The Normalized Planck curve is very useful. A lot of people are not aware that the Planck curve is a function of the single variable lambda.T, which is also a consequence of Wien's Displacement Law; lambdamax.T is a constant (2897.8 micron.Kelvins)

Also the peak spectral radiant emittance is 1.288 e-11. T^5 $W/m^2micron^{-1}$

Most people know that 25% of the BB spectrum radiation is emitted below the peak wavelength. The actual number is extremely close to 25%; but I have never actually done the integration to find out if that is exact; and if any of the PhD physicists out there know; they aren't telling anybody. But everybody should know that 25%:75% before and after the wavelength peak. And only 1% below 1/2 of the peak wavelength, and 1% left above 8 times the peak.

So it is kind of interesting that the solar spectrum long wavelength tail dies at about 4.0 microns, and the short wavelength edge of the Earth emitted LWIR also dies at about 4.0 microns; which is where the CO_2 peak that is dominant on Venus, sits. So that particular CO_2 band doesn't have much influence on Earth; well unless you are doing one of those fraudulent heat lamp CO_2 absorption demos.

I said:

> Every single additional molecule of H_2O or CO_2 that is added to the atmosphere WILL reduce the amount of incoming solar energy that reaches the surface of the planet to warm it.

And you responded:

> That assertion is not valid for CO_2, which is an absolutely-blind cat to short wavelength IR quantum/waves incoming from the Sun.
> —Nasif Nahle

Nasif, you are apparently determined to completely ignore the significant CO_2 absorption bands, at 1.8, 2.7 and 4.0 Microns wavelength which certainly address incoming solar spectrum energy. So I stand by my statement; but I anxiously await your contrary data that would show I am incorrect.

Ken Coffman: I am not claiming this has any significance when considering the mass of the Earth—but I found it interesting. Yes, I know this is contrary to my thoughts about viewing IR radiation as waves.

> *Light seems such a "flimsy" thing it is probably difficult to imagine it pushing on anything, or at any rate pushing very hard. And yet, the radiation force on the Earth due to the Sun is on the order of an impressive 100,000 tons! This is equivalent, for example, to a fully loaded aircraft carrier.*[144]

George E. Smith: The degenerate bending mode of CO_2 is the usual 15 micron band. The strongest CO_2 absorption band is the assymmetrical stretch mode, where all three atoms are in longitudinal motion along the axis of the CO_2 molecule. That is observed at 2345 cm^{-1} which is 4.26 microns wavelength. The symmetrical stretch mode where the C atom is stationary is predicted at about 6.3 microns; but is IR inactive since the dynamic dipole moment is zero.

Ken Coffman: Can we simply look at Dr. Pierrehumbert's first two sentences?

> *In a single second, Earth absorbs 1.22×10^{17} joules of energy from the Sun. Distributed uniformly over the mass of the planet, the absorbed energy would raise Earth's temperature to nearly 800,000K after a billion years, if Earth had no way of getting rid of it.*[145]

---

[144] Paul J. Nahin, *Oliver Heaviside, the Life, Work, and Times of an Electrical Genius of the Victorian Age*
[145] Raymond T. Pierrehumbert, *Infrared Radiation and Planetary Temperature, Physics Today*,
http://geosci.uchicago.edu/~rtp1/papers/PhysTodayRT2011.pdf

Beyond the silly, speculative nature of Dr. Pierrehumbert's thought experiment, he imagines an Earth at a much greater temperature than the source of the energy—the Sun. Maybe this is simply the pragmatic engineer in me, but I would never dream up an analogy like that. Nothing good can come of imagining this scenario—it's like a signpost on the road to madness.

George E. Smith: If ALL of the water on Earth was ice; then there would be no clouds and very little water vapor in the atmosphere, and suddenly we would have the full Sun beaming down on the surface at closer to the 1362 W/m^2, than 1000 W/m^2 like it does now. That would be the mother of all climate forcings, since even now, the Sun can heat the land surface to over 60 deg C during the day. With no clouds, there'd be no snow in the main land areas of the Earth; the boiling Sun would melt all that so it could run into the oceans and freeze.

What Peter Humbug is missing, is that water ice simply is not all that reflective. Fresh snow just a few minutes old may have 80% solar reflectance; but it quickly drops to a fraction of that once the Sun gets on it. Sea ice looks quite bright; but only when compared to sea water, which is near black (3% reflectance).

It is quite clear that CO_2 does virtually nothing to keep the Temperature up at nighttime in a high arid desert. Freeze all the oceans, and CO_2 won't do a damn thing about unfreezing them.

But the Sun beating down on the ice so the surface melts, and refreezes, creating an anechoic optical trap that conducts light and heat to great depths. And the Sun would evaporate all kinds of water vapor off the top of that ice even if there was not a single molecule of CO_2 in the atmosphere.

So you take out the CO_2 and you get a little bit less cloud cover globally; but I doubt that the global temperature would be perceptively different.

Our comfortable temperature range at today's orbital values, depends almost entirely on the Physical Properties of the H_2O molecule.

Ken Coffman

This of course could be disproven by simply publishing some peer reviewed, and experimentally observed data, that shows the global temperatures following the atmospheric CO_2; for any period of history, and any time offset, between the temperature and the CO_2. So far we've seen no such data.

Robert Clemenzi quotes me:

> So whatever your particles are emitting downwards, they must be emitting a like amount upwards; which escapes to space.
>
> Actually, the lower atmosphere is mostly IR opaque with a few windows. For most of the spectra were absorption occurs, 99% of the available energy is absorbed within 20 meters of the surface. In the "wings" it requires about 500 meters to absorb 99%. From this, I deduce that very little radiation from the lower troposphere ever gets to space. Based on my analysis of lapse rate plots, it is pretty obvious that, near the surface, the atmosphere is IR opaque and the net radiation is toward the surface. Since it is opaque, the radiation emitted toward space is simply reabsorbed within a few meters. At the tropopause, the atmosphere becomes IR transparent and a significant amount of energy is emitted into deep space.

If the Earth atmosphere is IR-opaque, then exactly how does LWIR from the upper atmosphere or even clouds ever reach the surface; and what is the directional selective mechanism, which causes the LWIR radiation to be concentrated downwards. The emissions from the surface are most definitely concentrated in the upwards direction. Emissions FROM the heated atmosphere itself are quite clearly isotropic; so they are not biased downwards.

If anything the escape path to space is favored over the downward path to the surface; simply based on the change in absorption line broadening due to Temperature and Pressure (collision broadening). Higher atmospheric layers are colder and less dense, so their absorption and emission lines are narrower than closer to the surface where the density (collision rate) and

Temperature (Doppler broadening) are increased.

There's no possible way the atmospheric radiated thermal spectrum can be concentrated downwards. The phenomenon is not optically different from the Raleigh scattering at short wavelengths; which makes the sky blue (daytime). The Raleigh scattering is also greater at ground level compared to the upper Troposphere, yet the blue sky looks the same upwards and downwards, except at the boundaries of the atmosphere (outer space, or the ground).

The Earth's atmosphere isn't even vaguely IR opaque; well not until you get out beyond any CO_2 bands and only H_2O is IR absorbing.

The surface emitted spectrum goes from about 5.0 microns to about 80 microns for 98% of the emitted energy, and CO_2 only grabs a small chunk of that in the 13.5 to 16.5 micron range. We are talking about a global "heating" source that has an average Temperature of 288K, and radiates about 390 W/m^2. An ordinary one pint (250cc) drinking water bottle is what is typically causing global warming via LWIR "heating."

Domenic: To Oliver Ramsay—Pierrehumbert completely downplays any effects of Nitrogen.

We live at the bottom of an ocean composed mainly of Nitrogen which in itself is a huge heat sink moderating temperature, and subject to all the fluid and thermal properties as would a liquid in an ocean.

1 atmosphere of pressure=10.3 m (34 ft) of water.

The atmosphere, mostly Nitrogen, is 'approximately' the equivalent of the Earth being covered in an additional 34 ft of water. Everywhere.

I don't think Nitrogen has ever been properly characterized for thermal radiation properties into the long wavelengths, let's say 8 microns and more. But that is absolutely necessary to characterize its insulating, or greenhouse properties.

I can't find any data. In the past, there was simply never any need to do it. Now there is. Both N_2 and O_2 need to be fully characterized for absorption, emittance, reflectance, and transmission in the long wavelengths.

You will find a lot on the short wavelength response of N_2 (O_2, CO_2, etc), but not long wavelengths. That is because short wavelength response is extremely easy to do. A high temp source is easy to build (a light bulb, hot plate, etc) and direct the thermal radiation through a room temp gas sample or send any other high energy particles through the gas sample.

The low temp characterizations have never been done because those are much more difficult measurements to make. You need a low temp sensor, as close to 3 or 4 Kelvin as possible to simulate outer space background thermal radiation, and then have the gas sample and a 290K or so radiation source emit through the gas, or vice versa.

Oops, I forgot to add:

The reason N_2 and O_2 need long wavelength characterizations is because there are MASSIVE amounts of it in our atmosphere. Even small absorption, reflection, etc amounts, because there is so much of it, will totally swamp any puny CO_2 effects.

These ivory chair pundits in academia need to get real....

Ken Coffman: No. I'm sorry. This following quote is as far as I'm reading. This man is not a scholar, he's a clown. That's not an insult, it's an observation. I suppose *Physics Today* is a peer-reviewed journal. Lovely.

> *An atmosphere is a mixed gas of matter and photons.*
> —*Dr. Raymond T. Pierrehumbert*

Oliver Ramsay: @Ken Coffman, if only he'd continued thus:

> *An atmosphere is a mixed gas of matter and photons, dreams and aspirations, airplanes and the rumble of history...*

he could have been a poet or philosopher (neo-post-mod).

_Jim: jae said:

> But, I would like to posit another "world" for the warmistas.
> Suppose the atmosphere of Earth consisted of ONLY N_2 and
> O_2. Now, these gases could not cool by IR emissions, but they
> would warm by conduction from the surface. And they could
> not radiate to space. So would they continually warm for
> millions of years?? Would they melt the planet? WTF, folks?

You are conveniently setting aside the sensible transfer of heat
energy to the poles via your N_2 and O_2, where, that surface cools by
IR radiation into 'space' (blackbody radiation proportional to T^4
power); I do not think you can deny that…

Domenic said:

> I don't think nitrogen has ever been properly characterized for
> thermal radiation properties into the long wavelengths, let's
> say 8 microns and more. But that is absolutely necessary to
> characterize its insulating, or greenhouse properties.
>   I can't find any data. In the past, there was simply never
> any need to do it. Now there is. Both N_2 and O_2 need to be
> fully characterized for absorption, emittance, reflectance, and
> transmission in the long wavelengths. …
>   Has that not been done already?

Infrared spectroscopy" N_2[146]

Domenic: To Jim: No.
    Either you did not read what I wrote, or you do not
understand it.

---

[146]

http://www.google.com/search?client=opera&rls=en&q=%22Infrared
+spectroscopy%22+N2&sourceid=opera&ie=utf-8&oe=utf-8

Spectroscopy is common. But NOT WITH LONG WAVELENGTHS as the source.

It should be done from 8 to 50 microns or so, if possible.

There is still a great deal of thermal energy being radiated at by the Earth at 17°C and lower surface temps. And a large net flow out because outer space is at 3 or 4 Kelvin.

You have to duplicate those conditions in a laboratory.

It is not easy to do.

David Ball: Thank you, George E. Smith for understanding what I am trying to say. AGW is in BIG trouble if it is shown that CO_2 follows temperature. The fact that we have had ice ages at much higher concentrations of CO_2 in the past is a big clue.

George E. Smith: @ Ken Coffman Well it seems that Peter Humbug is a bit iffy to me. He says Earth absorbs that amount of energy $1.22 \, E^{17}$ Watts (rate). Well it isn't going to get distributed uniformly over the entire Earth mass; and in addition the Earth DOES have a way to get rid of it; simple thermal radiation; and at 800,000K temperature, it is going to radiate at a much faster rate, that energy arrives from the Sun; so the Earth would cool, extremely rapidly; well it would never get to that temperature no matter how long you wait.

A rather childish example if you ask me; no—please DON'T ask me.

At lunchtime today at a Starbucks coffee shop, I talked with a high powered PhD Physicist. Make that a Particle Physicist—I stumbled across him recently when I noticed him reading a text book containing pictures that looked a whole lot like Feynman Diagrams. So I quipped; "You're not studying Particle Physics are you?"

"Well, actually yes," he replied.

He recommended that I read a small popular physics book *QED* by Richard P. Feynman, which I just received.

No QED does not mean "Quite Easily Done", or even the

Latin equivalent. It stands for Quantum Electrodynamics; the theory of the interaction between Photons, and Electrons, is how Feynmann puts it; and he was one of the primary instigators of QED at CalTech. It's the most thoroughly proven (by experiment) physics theory that we have; well almost anyhow; so I told him today that he probably needed QCD or Quantum Chromodynamics to do what he was doing. He seldom pokes his head outside the Nucleus; so QED is passé for him; but he agreed that QED was all I needed to completely understand molecular spectroscopy; and also to fully understand BB like thermal radiation. He fully agrees with me; that thermal radiation; that is Electromagnetic Radiation following Maxwell's equations and emitted from ALL materials at above absolute zero, including ALL gases; including N_2 and O_2 and even Ar; and he agrees that the fundamental Physical mechanism is simply accelerated electric charges following Maxwell's Equations for the EM field.

You see the molecules/atoms in a gas are zipping this way and that, and colliding with each other; and every time two molecules collide, they take off in some unknown direction with some unknown velocity; but with a Maxwell Boltzmann distribution of energies, and it is the impulse of that collision; and the "time of engagement" if I could use that term, to determine the acceleration that occurs for each colliding molecule; and it is that acceleration of the electric charge contained in the molecule that sets up the EM field that becomes the Thermal Radiation.

Now Maxwell's EM theory explanation came under fire with the postulation of the Bohr atom to explain the observed line spectra of gases. Maxwell insists that the electron running around in its Bohr atom orbit, must continuously radiate, so it would run down.

So Nils Bohr dismissed that objection with a quite unwarranted assertion that his orbital electrons did not radiate while they stayed in their orbit; but if they moved to a different orbit, then they radiated the energy difference as the E= h. nu radiation photon. Well the theory was a spectacular success in predicting the line spectra of Hydrogen and other atoms.

Now Bohr's "transgression" was not his insistence that the photon was emitted only when an orbit change took place; that part worked spectacularly. But you see he put the kibosh on Maxwell's insistence that accelerating charges must radiate; so Maxwell's equations looked like they were headed for the dust bin. And Bohr's assertion was totally without any supporting evidence; he just dreamed it up.

Well history salvaged both Maxwell's equations and also the Bohr Atom; or what it evolved into.

Bohr's planetary accelerating and radiating "Orbits" morphed into today's "Orbitals" or whatever we call them these days. Simply probability clouds mapping where the electron might be found, and the probability of it being there.

Voila!! Say Guv'nor, who said anything about the electron moving around; it's just there somewhere Mate; why would it move around?

Wonderful: if Bohr's energy-leveled Orbitals are just parking garages, and the electron is there somewhere and parked; it isn't accelerating, so it doesn't need to radiate per Maxwell's equations.

Well understand this is very much a stick in the desert island sandy beach sketch of what is going on. So QED to the rescue to explain it in ways that neither you nor I, nor even Feynmann can really understand.

So forget Bohr's planetary like Orbits; but NOT completely; the role they played in understanding Atomic structure and atomic spectra cannot be underestimated. The Bohr Atom IS one of the Crown Jewels of modern Physics; and Maxwell's equations survived it with little damage. Arnold Sommerfeld of course added much insight to Bohr's model; including the invention of the "fine structure constant" that has had a checkered life of its own.

So Maxwell survived, and accelerated charges DO radiate EM radiation, and at the molecular level, that manifests itself at ordinary temperatures in the LWIR emissions that we get form the Earth surface, and the oceans, and from the atmospheric gases themselves.

Keep in mind, that the capture of LWIR photons by GHGs, and the re-emission of thermal radiation from N_2 or O_2, or Ar, are continuous processes; so even though the gases have low thermal mass, and would cool rapidly on emission of a photon; they quickly pick up more energy from the collisions, as well as the continual capture by GHGs. So the atmosphere doesn't simply go chilly because it radiates; it is resupplied immediately in a continuous cyclic process.

And I freely invite any of the Physicist PhDs to polish off the rough edges of my somewhat sketchy tale here.

AJB: George E. Smith says:

> *Higher atmospheric layers are colder and less dense, so their absorption and emission lines are narrower than closer to the surface where the density (collision rate) and temperature (Doppler broadening) are increased.*

Whoaa—I was following along quite nicely to you said that. :-). Can you expand on this please? Correct me if I'm wrong; we have temperature rising/density falling as we ascend through the stratosphere then temperature falling and gradual physical stratification (with H_2O coming out on top) through the Mesosphere. How does all this fit together radiatively and what (if anything) is the apparent increased incidence of noctilucent clouds telling us?

John Whitman: George E. Smith says:

> *Keep in mind, that the capture of LWIR photons by GHGs, and the re-emission of thermal radiation from N_2 or O_2, or Ar, are continuous processes; so even though the gases have low thermal mass, and would cool rapidly on emission of a photon; they quickly pick up more energy from the collisions, as well as*

the continual capture by GHGs. So the atmosphere doesn't simply go chilly because it radiates; it is resupplied immediately in a continuous cyclic process.

George, I appreciate your focus and energy on these discussions.

The pieces are falling into place with the concept that N_2 and O_2 (and all other) gases in our atmosphere emit EM radiation in a continuous spectrum when they collide with any other molecule in the atmosphere, including with the spectral IR absorbing CO_2 molecule.

Now I am starting to appreciate much more what Tom Vonk was saying way back when during his post here at WUWT[147]

CO_2 does not appear to heat the atmosphere.

RICH: George E. Smith SAID:

Heat or energy in the form of photons or radiation travels in ALL directions including from cold to hot, otherwise **it would be impossible to see yourself in a mirror** *which is colder than you are.*

Isn't that due to reflection and not radiation?

Here you say:

And the second law of Thermodynamics as stated by Clausius, refers to **cyclic** *machines. "No cyclic machine can have no other effect, than to transport "HEAT" from a source at one Temperature, to a sink at a higher Temperature."* **doesn't say anything about EM radiation;** *just "HEAT".*

But here you say:

---

[147] WUWT, CO_2 heats the atmosphere...a counter view, http://wattsupwiththat.com/2010/08/05/co2-heats-the-atmosphere-a-counter-view/

> So the **atmosphere** doesn't simply go chilly because it
> radiates; it is resupplied immediately in a continuous **cyclic**
> process.

If our atmosphere is comprised of both EM radiation and heat, and
is cyclic, doesn't the 2nd law of thermodynamics rule our
atmosphere by default?

It seems to me that the heat reflecting properties of clouds,
working in conjunction with the 2nd law, is what creates our
atmospheric spin cycle.

CO_2 doesn't have reflective properties like a solid, does it?
Respectfully.

Ken Coffman: George, I imagine there was no discussion of
rarefied molecules storing IR, or slowing IR radiation's escape to
space (there can be delay, sure, but it's on the order of a
millisecond?), possessing amazing insulating properties or IR "back
radiation", all things AGW experts achieved consensus on…

Domenic: The pulse oximeter provides an excellent example of
light transmission. You can get an idea of how that works by simply
placing a tiny powerful flashlight on your fingertip. You will notice
how it 'glows' on the other side. Some visible light is penetrates
through the finger.

However, strictly speaking, the wavelengths they use are
visible, and very near visible, or near infrared, not what is
commonly known as 'infrared'. Notice they are using 650 to 950
nm wavelength light sources. That is simply .650 to .950
micrometers, or .65 to .95 microns.

Infrared wavelengths are roughly .70 to 1000 microns. When
the term 'long wavelength infrared' or 'far infrared' is used, it
means much, much longer wavelengths than those near the visible.

Now 'microwaves' wavelengths are much longer still than
even infrared: from 1000 microns to 1,000,000 microns, or rather

from 1000 microns to 1 meter in length!

In general, much of the stuff from the bulk of the infrared range will not penetrate the human body. It simply heats the surface. But microwaves can penetrate deeply into the body depending on the wavelengths used.

So, notice the situation: short wavelength visible light can penetrate the flesh to some extent as shown with the oximeter. Then most of infrared cannot. Then microwaves can, etc. Do you see the pattern? Short wavelengths penetrate. Medium do not. Then longer still do penetrate. etc, etc.

'Transmissivity' through materials is wavelength dependent. So is emission, reflection, absorption. It is ALL material specific, depending on the atom and/or molecule.

To make it all even more complicated it is also ANGULARLY dependent, depending on the angle that the thermal radiation strikes the atom or molecule. Some atoms and molecules are very reflective of IR at certain angles of incidence, and highly absorptive at other angles of incidence. But you probably have enough to chew on for now.

Oliver Ramsay: @ John Whitman, Ditto on George's contributions.

I can't agree about the Tom Vonk thread. I was convinced that it was all just a question of semantics and that nobody quarreled with CO_2 absorbing some energy and passing it on to the other gases.

All the craziness built on that premise is what keeps us entertained.

Domenic: To George E. Smith, yes, now you are getting it. Now, also in regards to Peter Humbug's (great play on the name!) paper. The angular effects from the energy of the Sun striking the Earth are very significant.

When the Sun is directly overhead, the absorption by water of the visible components of the total energy are MUCH greater than

when the Sun is near the horizon. You can 'see' that for yourself if you look out over the ocean at noon and then at dawn or dusk on a clear day. At noon, you will see almost no reflection of the Sun on the water. Some of it, of course is reflected straight back upwards skyward, but a lot of it simply penetrates straight into the water.
In the tropics, the Sun spends a great deal of time in the overhead position, penetrating water deeply.

In the polar regions, however, the Sun only goes up to a very narrow angle over the horizon. Hence, its ability to penetrate the polar water mass and ice mass with thermal energy is GREATLY diminished. Much visible solar energy that could be absorbed is NOT. It is simply reflected away back into space because of the narrow angle. (Snow blindness, etc.)

So, there is a HUGE difference from solar effects on the tropics compared to the polar regions.

Oliver Ramsay: @Domenic, that is very reasonable for planet Earth, but it's clearly different on a planet with a temperature of 288K at all locations, at all times.

It seems that, although mention is made of latitudes and hemispheres and even diurnal variation, the planet AGW is an accretion of averages, with Earth-like phenomena painted on the surface for decoration.

This why the equator will be glaciated if we don't get the whole CO_2 elimination trick just right.

Domenic: To Oliver Ramsay, they do indeed live on a different planet.

Those Peter Humbug types who simply hang around their comfy little offices and other artificial environments. To them, it must be easy to envision a uniform 288K world.

But it's not the Earth.

It's a figment of their imagination.

It's also called delusion.

NoIdea: @ Ken Coffman, thank you for mentioning the ridiculous nature of the humbug paper's first paragraph, it does indeed make reading the rest of it feel like a waste of life.

@ Oliver, thank you for the inspirational line, I allowed a few to follow it, just to see where it went—I name it:

HELIOCENTRIC HUMBUGGERY

An atmosphere is a mixed gas of matter and photons, dreams
and aspirations, airplanes and the rumble of history. The
blogosphere is a mixed matter of gas and phlogiston, nightmares
and fears, chem-trails and the neon visions of the future.
Neo post modernistic philosophers search for the stone
unturned, the poet for the phrase yet learned but still realistic.
Warmest warmonger alarmists still shill; shrieking and striking
at all they do not understand and cannot tell.
Listening to the voices only they can hear, turning the light of
truth aside, no reason to reason only to hide.
In dark terror despising all thoughts arising, potentially
disguising for a brief moment their hatred of life.
Do these "experts" really believe that if you add 2 pans of
water that are both 95°C they will become 190°C?
Have they never heard of laws that stand unmolested, standing
the tests of time and engineering our life comforts?
Shading our eyes from an incandescent source, our fingers held
tight but glowing with the light seemingly flowing.
So many everyday proofs that what the fearful refuse to
believe, it is out of our hands and into the face of the Sun that we
stare.
Blinded by the force of authority, no room for a thought, no
time for a dream, just time for terror and voice to silently scream.
Taking out the extraneous, adding in the superfluous,
searching for the essence of what lays hidden in plain vision.
Poetry is the mathematics of a language, weird as that sounds.
Minimal enough to be nothing but exactly zero and still be found.

Sometimes a wave is as good as a particle to a blind bat, the
spiraling nature of insanity is lost on the long winded fans.
Cabalistic and cannibalistic biting the hand that feeds them, the
leeches launch caution into the path of the principled precautionary
Humbuggery overwhelmed peers of consensual consensus, pal-
reviewed policies of global genocide so reckless.

Bill Illis: George E. Smith says:

> ...*thermal radiation; that is Electromagnetic Radiation
> following Maxwell's equations are emitted from ALL materials
> at above absolute zero, including ALL gases; including N_2 and
> O_2 and even Ar; and he agrees that the fundamental Physical
> mechanism is simply accelerated electric charges following
> Maxwell's Equations for the EM field.*
>
> *You see the molecules/atoms in a gas are zipping this
> way and that, and colliding with each other; and every time
> two molecules collide...*

At sea level pressure, each atmospheric molecule collides with
another molecule 6.7 billion times per second. A rate that is 33,000
times faster than the average emission time for CO_2. So, an excited
CO_2 molecule (which is now effectively many times warmer than
the surrounding air) crashes into another atmospheric molecule
33,000 times before it can emit. In the troposphere, the numbers
are not much different, 10,000 times.

So, if N_2 and O_2 and Argon are actually emitters and absorbers
of IR (albeit at a far reduced level than the specific absorption and
emission frequencies of CO_2), and are crashing into CO_2 at a far
faster rate than CO_2 emits at, the entire picture of atmospheric
radiation changes.

There is less emission at the CO_2 and H_2O specific absorption
frequencies (because it is getting siphoned off by all the other
molecules before being emitted) and, instead there will be
increased emissions in the blackbody spectrum of the Earth's

effective temperature where N_2 and O_2 are operating like blackbodies.

And that is exactly what the emission spectrum of Earth shows. The effective emission temperature in the CO_2 bands is only 220K. While the blackbody atmospheric windows are 290K (while the average is supposed to reflect 255K).

This is a different perspective than is commonly explained in the greenhouse radiation theory.

It's just that you cannot find anywhere on the internet, someone that shows N_2 and O_2 and Argon absorb any IR at all. Everyone seems to believe/has been taught, that N_2 and O_2 do not absorb any IR at all, even blackbody-type radiation.

barn E. rubble: I appreciate the efforts of all those providing information &/or opinion that has been presented here, very much.

From what I've read here I'm finding even harder to believe that 'heat' &/or 'heat energy' or even just 'energy' could be transferred to the deep ocean depths >900m. Trenberth has lately claimed that that's where he'll find the 'missing heat' re: energy balance. This transfer was apparently undetected during the transfer and remains undetectable. From what I understand Trenberth believes ALL current and previous temp. data are wrong, as well as all interpretations of the available data.

Is it possible for any IR energy to penetrate >900m? Further could that happen without being detected with current technology?

Again, thank you for the time and effort of teaching...

Mikael Cronholm: @ barn E. rubble. "Is it possible for any IR energy to penetrate >900m?" Within the wavelengths of IR cameras, 2-14µm: Definitely NO! And unless we can identify magic IR wavelength bands that penetrate everything, I seriously doubt that there is much radiation in any E/M band that reaches down there. As a SCUBA diver I know that the longer wavelengths of visual begin to disappear pretty fast, even at the depths of amateur SCUBA, red start to disappear and things look more blue. I have heard of fish that live at extreme depths having red color as camouflage. There is no red light there anymore, so nothing will reflect off them, and they "disappear."

I don't remember the exact numbers now, but I have read some time that the first 10m or so of the oceans contain the majority of the heat stored in the oceans.

@ Will. On the site you refer to with your link they say, under "Infrared Windows": "An IR transparent weatherproof shield or window can be made of polyethylene which is transparent to IR radiation in the 5~15 micrometer wavelength range." They suggest PE, because glass is opaque. The lenses they use are plastic, because glass is opaque.

And here something for you:

> An example of a material whose emissivity characteristics change radically with wavelength is glass. Soda-lime glass is an example of a material which drastically changes its emissivity characteristics with wavelength (Figure 2-5). At wavelengths below about 2.6 microns, the glass is highly transparent and the emissivity is nearly zero. Beyond 2.6 microns, the glass becomes increasingly more opaque. Beyond 4 microns, the glass is completely opaque and the emissivity is above 0.97.

And here you can find that quote, and a nice spectral graph (Figure 2-5 above) that describes how the transmission changes with wavelength. It even shows different thicknesses and how that changes the transmission.[148]

And with that, I am done with the discussion about the IR transmission of glass. I could show you a live demo here at my office, but you would have to pay the ticket to Thailand yourself.

---

[148]

http://www.omega.com/literature/transactions/volume1/theoretical3.html

Domenic: To Will, you are not reading what Mikael wrote very CAREFULLY. Mikael wrote that common glass starts decreasing transmission at 2 microns, and at 2.3 microns stops nearly all further transmissions.

So, yes indeed, glass does transmit 'some' IR, BUT ONLY A TINY AMOUNT compared to the entire infrared band!

Infrared originally simply meant 'beyond the color red" because of Herschel's discovery in 1800. However, later it was defined as the region of 0.7 to 1000 micron wavelengths. Re: the pyroelectric motion detector

First of all, those are the cheapest IR systems you can buy. Any specs they give should be treated as 'suggestions' rather than 'specs'. Most are built in China. The quality of the optical coating on the internal silicon window they use is the cheapest possible.

If you explain to me EXACTLY what you did with that glass and al foil in your test of that pyroelectric security sensor, I will tell you EXACTLY where your error in understanding is.

You cannot compare that cheap pyroelectric device with the equipment that Mikael uses. Mikael's equipment is much more sensitive and precise, but it also has its limitations.

Regarding window coatings, you do not yet understand the concepts of emissivity and reflectivity. If you did, you would see where your reasoning has failed.

Regarding O_2 and N_2, that part, I agree with you strongly. They have not been characterized fully for absorption, reflection, transmission, and emission over the ENTIRE infrared band 0.7 to 1000 microns. That is one the major flaws in the way the warmists have been defining greenhouse effect. I have been pointing that out in multiple posts here.

Robert Clemenzi: Response to George E. Smith...
There are several frequency bands where the atmosphere really is IR opaque. Even though radiation is emitted in all directions, it is reabsorbed within 10 meters or so. Therefore, the fact that the line widths change with temperature and pressure does not have an

effect. As a direct result, more energy is returned to the surface from the warm lower layers than released to space from the cold upper layers.

As for clouds, these are blackbody emitters and their energy passes through the windows where greenhouse gases don't absorb. Except for a very small amount of energy from ozone, no IR energy from the upper atmosphere reaches the surface. None.

Your claim that "the escape path to space [should be] favored over the downward path" would be true if the atmosphere was only partly absorbing **at those frequencies**. To be clear, the atmosphere is IR opaque in most of those frequencies where water vapor and CO_2 absorb, and IR transparent in the "windows".

barn E. rubble: Do I understand correctly that CO_2 levels, however increased, does not in fact mix or average out at increased levels globally but would accumulate locally, i.e., where it is being produced and cycle about unless moved by other forces and dissipate sooner rather than later?

Further, the GH effect of CO_2 as per warming the surface temp by re-emitting IR/heat energy towards the surface is negligible? Are there no peer-reviewed papers (for what that's arguably worth) on the natural separation of in our atmosphere? Growing ever so skeptical, it would appear the main plank in supporting AGW; well established physics, RE: CO_2, seems to be—well, less so? Leaving the only other plank; consensus, somewhat shaky?

Excuse me for taking the discussion back a few grades for just a moment. My understanding of AGW simply put: Energy from the Sun comes in and hits the surface of the Earth warming it. Energy is emitted back from the Earth where CO_2 (among others) blocks/absorbs and re-emits energy back to the surface. Increasing CO_2 levels will enhance this by blocking/absorbing more energy and thereby re-emitting more energy/heat back to the surface. And so on, until the dreaded tipping point. Apparently that means our atmosphere must surely be getting bigger as well, like blowing up a balloon, if that is, there isn't an equal (and natural) release somewhere. Does there not have to be a volume balance to deal with and not just the energy balance for AGW to work? Considering the panic of increased levels of one component should there not be an equal panic (at least a question or two?) about what components are being displaced?

Back to IR:

A small aside here from examples used earlier on. There is no such thing as an IR sauna. Full stop. Perhaps just a misnomer but the literal translation of 'sauna' is steam bath. You can no more have a 'dry' sauna than you can have a 'dry' shower or bubble bath. Call it a warm room or sweat closet but a sauna it is not (pronounced: sow-na) it irritates your Finn friends, believe me. Always sweating the small stuff…

George E. Smith: Mikael Cronholm says @ barn E. rubble.

> *Is it possible for any IR energy to penetrate >900m?" Within the wavelengths of IR cameras, 2-14μm: Definitely NO! And unless there are some magic IR wavelength bands that penetrates everything, I seriously doubt that there is much radiation in any E/M band that reaches down there.*

Sea water at its most transparent point; which occurs at about 460-70nm wavelength (blue), has an absorption coefficient of between 1 and 2×10^{-4} cm^{-1}. So that means that that wavelength will be attenuated down to $1/e$ (37%) in a depth of 10^4 cm max, or 50-100 metres for that range (I can't read the logarithmic scale on the graph any more accurately). So five absorption lengths will reduce the residual down to 1%, which is 250 -500 metres. For 5% remaining you go three times, so 150-300 metres; and that is for the MOST penetrating wavelength. At the UV end of the visible spectrum, (380 nm) the absorption is 10 times as much; 0.001 cm^{-1}, so 99% extinction in 50 metres depth.

For the red end (780nm) the absorption is another ten times

higher, or 0.01 cm^{-1}, so 99% is gone in 5 metres of good clean clear sea water, uncluttered with plankton etc.

At 1.5 microns wavelength in the near IR, you run into a significant water absorption band, where alpha is 30 cm^{-1}, so the 1/e depth is 333 microns, or 1.6 mm for 99% loss. There's a slightly higher peak at 2.0 microns, and then the highest of all at 3.0 microns where alpha is 8-9,000 cm^{-1} or 1.1 to 1.25 micron absorption depth. Longer than that, there's some dips, but not below 100 cm^{-1} for alpha, and after 7.0 microns, it settles down to around 1000 cm^{-1}, at least out to 100 microns wavelength which is about the limit of our interest for climate issues. Then water slowly gets more transparent about as the square root of the wavelength (or frequency) all the way into the radio spectrum. It finally gets back to around 1 cm^{-1} at around three metres wavelength (100 MHz).

So no; there is no part of the EM spectrum, of concern for climate issues, that can penetrate to 900 metres depth in the ocean.

Those data are from two graphs in the same paper.[149]

Michael H Anderson: At the end of the day, the salient point is really in those last 3 paragraphs.

If not, it isn't science, it's guessing.

Kind of obvious really, but elegant in the way such quotes often are.

Mikael Cronholm: Emissivity and absorptivity is essentially the total opposite of each other, emissivity tells you how well something 'gives off' radiation and absorptivity how well it 'takes up and retains' radiation. But here is the clue, and it is a very fundamental thing and basic in radiation science. At any given wavelength and angle of incidence the absorptivity and emissivity of

---

[149] G.C. Ewing, *Oceanography from Space*, Woods Hole Oceanographic Institution, Woods Hole Mass WHOI ref.No 65-10 April 1965

a surface will ALWAYS be the SAME VALUE. This is called Kirchhoff's Law, but if you Google it make sure don't end up with the one that deals with electrical circuits.

That law is fundamental, because without it we could not achieve equilibrium in a system. So, if something has an emissivity of 0.97, the absorptivity is exactly the same.

Radiation equations:

$$a + r + t = 1$$
$$e = a$$
$$e + r + t = 1$$

For most objects we can assume $t = 0$, so:

$$a + r = 1$$
$$e + r = 1$$

And I would still like to know why a window would look black in the IR camera because it is opaque?

Sorry, I see now that you have answered that already. So with your infrared camera, have you ever looked at windows or glass with different temperatures? And, just a question, what wavelength is your camera and what is normally your purpose for using it?

George E. Smith: Robert Clemenzi says in response to George E. Smith:

> There are several frequency bands where the atmosphere really is IR opaque. Even though radiation is emitted in all directions, it is reabsorbed within 10 meters or so. Therefore, the fact that the line widths change with temperature and pressure does not have an effect. As a direct result, more energy is returned to the surface from the warm lower layers than released to space from the cold upper layers.
>
> As for clouds, these are blackbody emitters and their energy passes through the windows where greenhouse gases

don't absorb. Except for a very small amount of energy from ozone, no IR energy from the upper atmosphere reaches the surface. None.

Your claim that "the escape path to space [should be] favored over the downward path" would be true if the atmosphere was only partly absorbing at those frequencies. To be clear, the atmosphere is IR opaque in most of those frequencies where water vapor and CO_2 absorb, and IR transparent in the "windows".

What's to understand? The mean temperature of the Earth (surface or Lower Troposphere) is purported to be 59 deg F or 15 deg C; 288 Kelvins. So a blackbody at that temperature radiates a spectrum that peaks at 10.1 microns, and contains 98% of its total energy emissions between 5.0 microns and 80 microns, with only 1% beyond each end. Now temperature extremes range from as low as 183K around Vostok to as high as 333K in the tropical deserts (surface).

So that will extend the spectrum somewhat. But we should note that NO Black body emits MORE energy at ANY wavelength, than is emitted from a black body at a HIGHER temperature. So we can completely discount any areas that are substantially colder than the global average; their emissions don't show up as any perceptible increase in the global average.

The high Temperature limit (333K) will drop the spectral peak down to around 8.8 microns, so extending the low wavelength limit to say 4.4 microns.

So we can say that the LWIR spectrum of the Earth covers perhaps 4.0 to 80 microns wavelength range.

Of that, CO_2 carves out a notch from about 13.5 to 16.5 microns. At least half of the entire spectrum energy is emitted entirely below that 13.5 micron band edge of CO_2; and at least 35%, maybe as much as 40% is entirely above the 16.5 micron upper edge of the CO_2 band. And those higher Temperature tropical desert areas are emitting as much as 1.8 to 2.0 times the global mean radiant emittance; and that into a nearly water free atmosphere, and at wavelengths even further removed from the CO_2 band; so even more escapes.

That leaves at MOST, 10-15% of the entire Earth LWIR Radiant energy emissions subject to CO_2 absorption.

The rest of that energy, with dry air, is pretty much free to leave. Please don't insult us here at WUWT by claiming that H_2O is a greenhouse gas; the AGW folks are most strident in their assertions that H_2O is merely a feedback amplifier for CO_2 the Prince of green house gases.

Absent H_2O, the atmosphere isn't even vaguely IR opaque. But back to our water bottle. As I said, the Earth LWIR spectrum is reasonably from about 4.0 to 80 microns with a peak near 10 microns; and corresponds to a black body thermal radiation source (roughly) at a temperature of around 288K; 300K or whatever; that is the temperature of a source for the radiation that is involved in the greenhouse effect (which most WUWT readers readily acknowledge).

Now water has a reflection coefficient of about 3% tops, so a bottle of water at about room temperature is perfectly good laboratory source for the kind of LWIR radiation, that is absorbed by CO_2 and causes the greenhouse effect.

100 Watt light bulbs are NOT a good source for Earth like LWIR thermal radiation. The wavelength is 10 times too short, and the Brightness is 10,000 times too large, and the temperature is 10 times too high.

> There are several frequency bands where the atmosphere really is IR opaque. Even though radiation is emitted in all directions, it is reabsorbed within 10 meters or so. Therefore, the fact that the line widths change with temperature and pressure does not have an effect. As a direct result, more energy is returned to the surface from the warm lower layers than released to space from the cold upper layers.

When you say it that fast; you can almost get people to believe that. But that assertion is based on a popular fallacy; the notion that gases do NOT radiate a thermal spectrum based on their temperature, so the "re-emission", is the same 13.5 to 16.5 micron CO_2 absorption spectrum. And that is clearly not the case.

The atmosphere clearly does radiate a normal temperature based thermal Spectrum, and yes it does contain some absorption dips notably a narrow one around 9.6 microns, from the thin high level Ozone layer, and the 13.5 to 16.5 micron CO_2 band; heck, take 13 to 17 microns if you like. The point is that the bulk of the thermal radiation from the atmosphere and from the surface is NOT within the CO_2 absorption band, so it can proceed to space through a dry atmosphere, with just a small recapture by CO_2. What the CO_2 doesn't recapture; which is most of the spectrum, escapes; subject ONLY to the water conditions; and we know that water is not a GHG; just a CO_2 helper.

My source for the atmosphere radiation data is: H. Rose, et al.[150]

Yes it is true that in the near vacuum of the stratosphere, the mean free paths are long enough for CO_2 to spontaneously re-radiate its 13.5 to 16.5 micron band; but in the lower levels (you mentioned the 10 metres or so near the surface), the captured energy is completely thermalized through molecular collisions; so the source of the atmospheric LWIR is ordinary thermal emission form the ordinary atmospheric gases, and that emission is entirely independent of any trace GHG content; and depends ONLY on the temperature of the atmosphere.

And if you really want to permit H_2O to take its proper place in the mechanics of the atmospheric energy (radiant) processes; then you will quickly discover how inconsequential CO_2 really is.

Since H_2O is the ONLY condensing GHG, it is the only one that can form clouds, which immediately introduce a huge net

---

[150] *The Handbook of Albedo, and Thermal Earthshine*; Environmental Research Institute of Michigan (ERIM), Ann Arbor, MI, Report No. 190201-1-T

cooling effect. Nobody EVER observed it to warm up (go to a higher temperature) in the shadow zone, when a cloud moves in front of the Sun; it ALWAYS cools down.

> ...no IR energy from the upper atmosphere reaches the surface. None. Your claim that "the escape path to space [should be] favored over the downward path" would be true...

Somewhat conflicting aren't these notions?

Domenic: to Barn, the most pristine data sets to relate these IR discusssions to are those from Amundsen Scott AFB and Vostok bases in Antarctica.

The Sun is a minor effect there. Temperatures are almost completely dominated by radiational heat transfer to outer space through the atmosphere.

Outer space is a constant. Nighttime sky does not vary as a blackbody target.

If CO_2 was acting as a powerful greenhouse gas, it would show up in their data first, as CO_2 increases in the atmosphere there would trap more heat radiationally.

But it is not.[151]

And the data would be clearer if someone went and picked out simply the lowest recorded temperature per year at those two locations. And then graph those.

You see, there is absolutely nothing on Earth that can drive those temperatures lower.

Nothing.

Except the 'greenhouse effect' lessening.

George E. Smith: Bill Illis says:

> ...thermal radiation; that is Electromagnetic Radiation

---

[151] http://icecap.us/images/uploads/VOSTOK.pdf

following Maxwell's equations are emitted from ALL materials at above absolute zero, including ALL gases; including N₂ and O₂ and even Ar; and he agrees that the fundamental Physical mechanism is simply accelerated electric charges following Maxwell's Equations for the EM field.

You see the molecules/atoms in a gas are zipping this way and that, and colliding with each other; and every time two molecules collide,

At sea level pressure, each atmospheric molecule collides with another molecule 6.7 billion times per second. A rate that is 33,000 times faster than the average emission time for CO_2. So, an excited CO_2 molecule (which is now effectively many times warmer than the surrounding air) crashes into another atmospheric molecule 33,000 times before it can emit. In the troposphere, the numbers are not much different, 10,000 times.

So, if N₂ and O₂ and Argon are actually emitters and absorbers of IR (albeit at a far reduced level than the specific absorption and emission frequencies of CO_2), and are crashing into CO_2 at a far faster rate than CO_2 emits at, the entire picture of atmospheric radiation changes.

Bill, YOU get it, I get it, some others get it. Why is it so hard for others to grasp?

The LWIR radiant energy that is specific to the GHG absorption bands (13.5-16.5 microns for CO_2), is what the GHG molecules CAN and DO capture; from the entire 4.0-80 micron LWIR spectrum corresponding to Earth temperatures; and that capture energy is immediately thermalized (thanks for coming up with the numbers); and as of that point; the GHG molecule is totally removed from the picture (well to capture again).

And even though N₂ and O₂ or Ar, and other non dipolar molecules do not capture IR through resonance absorption bands, as do CO_2 and H_2O, those air molecules are still quite free to radiate a perfectly normal thermal emission spectrum, that depends ONLY on the atmospheric temperature, and is quite insensitive to the mechanism that heated the atmosphere in the first place.

That heating could have been incoming solar heating, or ground contact conduction and convection; by whatever means the atmosphere can be warmed (including deposit of latent heat from water condensation (or freezing); it will then radiate according to thermal radiation laws; and MOST of that thermal radiation spectrum is NOT subject to further assault by CO_2; more is by H_2O though.

As Phil several times pointed out here, in the stratosphere mean free paths can be long enough for GHGs to spontaneously decay to the ground state.

Even though the atmospheric gases are low thermal mass, and must therefore cool from radiating thermal emission; they are maintained at the ambient temperature by immediate resupply from the molecular collision processes; so it is a continuous energy pumping process.

Oliver Ramsay: If I were a fotong from the stellar entity Sol who had somehow been cast abruptly down to the Earth's surface, I would be trying to get back to the mother ship as quickly as I could. Falling into the clutches of CO_2 and having to ping-pong my way back to freedom through the troposphere, I would be delighted to be back-radiated to the ground 'cos then I'd have another chance at a clear shot through the window.

Getting my bearings after that might prove difficult but, if I did make it back to Sol, I would make my home "warmer than it otherwise would have been".

Mikael Cronholm: I have previously only had a casual interest in the issue of global warming (or not?) and I still don't know what the final answer may be, but I have gained some very valuable insights from many of the comments made here.

On the definition of heat, we probably don't even need to

agree to disagree, since we are both victims of the semantic confusion that exists. I no longer make the mistake of arguing one way or the other, and the choice of principle I have made for teaching is for purely didactic reasons. The physics remain the same anyway!

Domenic: Hi Mikael, it's been very enjoyable to join you in these discussions. I really hadn't engaged much in the AGW debate until lately. There really hadn't been any good venue to be able to do so until Anthony got WUWT up and running.

It's been about seven years since I've even engaged in any 'infrared' discussions in any meaningful way with interested parties. So, it has been a good exercise to share in that. And doing so, has triggered some insights that I hadn't seen before.

I don't mind the 'Will's of the world. The science of infrared, and the thinking involved, is quite arcane and alien to most people. By its nature, it forces one to look at 'wholes' rather than 'pieces' as is common in the thinking of most people.

I have noted your email, thank you. I don't think I will be getting to Thailand for at least a few years, but I do appreciate it.

Robert Clemenzi:
http://www.coe.ou.edu/sserg/web/Results/Spectrum/o2.pdf
and
http://www.coe.ou.edu/sserg/web/Results/Spectrum/n2.pdf
show the IR spectra of O_2 and N_2, respectively. Unfortunately, there are no spectra for H_2O and CO_2 for comparison. The following table compares the peak absorption coefficients per molecule for several molecules.

CO_2 1 E-19
H_2O 1 E-18
O_2 1 E-28
N_2 1 E-28

The bigger the value, the more energy is absorbed. Granted, these numbers must be multiplied by the total number of molecules in the column to get the actual absorption. However, the relative concentrations can be used to make simple comparisons. Since water vapor is about 1% of the lower atmosphere (10,000PPM) on a dry day and because it absorbs at many more frequencies than any of the other 3, you can see why it is the main greenhouse gas in the troposphere. Basically, after adjusting for the relative concentrations, it is 8 orders of magnitude stronger than O_2 or N_2. At the bottom of the tropopause, water vapor is still about 70PPM and it is still the strongest greenhouse gas.

At any rate, these numbers explain why so many people agree that O_2 and N_2 are IR transparent.

In addition, for a given molecule, radiation is absorbed and emitted at exactly the same frequencies (same spectrum). It is not correct to assume that O_2 absorbs with one spectrum and emits with a blackbody spectrum. The only exceptions I know to this are fluorescence and phosphorescence.

For George E. Smith: The radiation spectra for a gas at a given temperature is the blackbody spectra for that temperature times the spectra coefficients from the HITRAN database (adjusted for temperature and pressure) times the number of molecules in the path. Thus the emission spectra is always less than or equal to the blackbody spectra.

Also, you seem to think that desert air contains no water. Actually, it contains a lot (perhaps 1,000PPM or so), just not enough condense and form dew and/or fog in the mornings.

Mikael Cronholm: @ Robert Clemenzi. As a matter of fact, even desert air will condense, but not on the ground, which is perhaps what you mean. The ground has too high heat capacity to cool down enough by radiating heat to the sky. But on objects with less heat capacity, water will condense from the air. That is how a cactus can survive. Water condenses on the needles, or barbs, or what you call them, because they are dry in themselves and away from the cactus that remains warm due to the heat it has accumulated. The condensed water on the needles will run down to the roots and be absorbed there. Cactii have short roots and with very sporadic rainfall they will have no other way to reach water except that the cactus makes the water itself.

In military special forces they use old desert techniques to create water from the air by placing a piece of cloth or something on sticks with a small rock in the middle to make an upside down tent-like device which will cool down towards the sky, enough to go below the dew point and get condensation.

It can be REALLY cold in the desert at night too. Because of the relative lack of greenhouse effect, due to low levels of water vapor. Which kind of ties this together with the topic :-)

Bill Illis: @Robert Clemenzi
Again, this is based on the assumption that N_2 and O_2 are zero (on the scale of 0 to 1 for blackbodies).

We know N_2 and O_2 warm up to reflect the air temperature at whatever location or height they are in the atmosphere. So no emission or absorption other than these two very weak specific spectra?

From Raymond Pierrehumbert's recent article on planetary atmospheres and the greenhouse effect (Pierrehumbert is the newest star expert on atmospheric physics on the pro-AGW side having also written a long textbook on the matter which will likely become a key source at universities etc.)

An IR photon absorbed by a molecule knocks the molecule into a higher-energy quantum state. Those states have very long lifetimes, characterized by the spectroscopically measurable Einstein A coefficient. For example, for the CO_2 transitions that are most significant in the thermal IR, the lifetimes tend to range from a few milliseconds to a few tenths of a second. In contrast, the typical time between collisions for, say, a nitrogen-dominated atmosphere at a pressure of 104 Pa and *temperature of 250K is well under 10^{-7} s. Therefore, the energy of the photon will almost always be assimilated by collisions into the general energy pool of the matter and establish a new Maxwell–Boltzmann distribution at a slightly higher temperature. That is how radiation heats matter in the LTE limit.*

He then goes on to explain the formulae etc. and then writes:

We can determine that temperatures of the atmosphere and ground range at least from 220K to 285K. But absent additional information, we cannot tell that the high end of that range actually comes from the ground.

The 220K is where the CO_2 spectral band dominates and 285K is the atmospheric windows where N_2 and O_2 blackbody radiation dominate. But they cannot tell what the origination point is for the 285K blackbody radiation from N_2 and O_2. It might be directly from the ground or it could be 20 kms high where the density of air falls to a level which starts to allow a 50:50 chance of a photon being emitted directly to space.

I choose to believe the "lapse rate" tells us exactly what is happening to the energy levels in the atmosphere and where the radiation is going and which molecules are excited etc.

It is different than the way Pierrehumbert explains it and it does not completely ignore blackbody radiation which seems to be how the global warming theory got started—by focusing on the specific absorption bands only and ignoring collision rates and blackbody radiation in the atmosphere.

Ken Coffman: Can I add one more dumb question? Let's take an N_2 molecule as an example. I think we all agree—if it has a temperature, it will radiate and for N_2 the wavelength is real short. Nonetheless, isn't this emitted energy highly-significant from an energy balance POV? It's a short wavelength, but isn't there a lot of it? I feel like there is a missing piece in my thinking.

Thanks!

Domenic: To Robert Clemenzi, unfortunately, those curves you link to cannot be trusted. They are only partial data.

Notice the preponderence of peaks up to 6 microns or so, then the lack of any peaks for wavelengths longer than that. In the O_2 graph, for example, they skip all the way to nearly 1000 microns. They do that in the graphs because real data has never been taken in the 6 to 1000 micron range. They can only put in what they have tested or guessed (calculated).

That is the problem.

If you have followed my posts here, I have explained why. It is because those physical measurements are difficult to make. But in order to understand the true effects of O_2 and N_2 in the atmosphere and their contribution to the greenhouse effect, those physical absorption measurements must be made. Real data must be taken in those long wavelengths, not guessed at or calculated.

To Ken Coffman, and others still here, because of this topic, I've started to look very closely at the Antarctic Vostock and Amendsun-Scott data. It's the most pristine data out there. You want to know what is really strange—that the data I've seen so far, from a radiational physics point of view, is showing that the 'greenhouse effect' of all the gases in the atmosphere is actually DECREASING! That surprised me.

Despite all the stuff we supposedly are putting into the atmosphere. Now that is indeed a shocker!

But then "Nature" has always been much different than many 'think' it is, always confounding man....

I am going to look into it more thoroughly. In addition, it is water vapor, H_2O, that has been saving our butts.

barn E. rubble: This has been quite enlightening and entertaining. Thank you, to all those who have offered their expertise, knowledge and opinion. I appreciate the patience shown to those of us PhD challenged. I'm hoping all those contributing on this thread haven't signed off yet because I was looking for a brief summary regarding IR and the AGW theory that I can take away from all of this.

Two points from the thread (and unfortunately I didn't copy contributor names).

> CO_2 negligible to global warming because: "That leaves at MOST, 10-15% of the entire Earth LWIR Radiant energy emissions subject to CO_2 absorption." i.e., CO_2 only absorbs (and re-emits in 3D, Earth, atmosphere and space) a fraction of IR/heat energy that is being referred to as 'back radiation', the main issue driving AGW.

> CO_2 significant in global warming because: "The simplest and quite accurate way to conceptualize this is that CO_2 is an insulator..." i.e., CO_2 absorbs (and re-emits back to Earth) a significant amount of IR/heat energy that is being referred to as 'back radiation', the main issue driving AGW.

Have I got the main issues here? Thank you for your time and trouble.

Regarding IR with respect to O isotopes, perhaps someone here can shed some light or even a thought on the following. Somewhere on this thread someone referred to some charts with Oxygen isotopes and corresponding IR wavelengths. I have been reading with some interest of the work coming out of the Univ. of Saskatchewan Isotope Lab where they've been using O isotopes for temperature reconstructions. The colder the temp, the more heavier O (18?) was found.

The question is: how does the IR/heat energy cause more (or less) of one isotope than another, i.e., does temperature change the isotopes?

305

Michael H Anderson: Oops: Climategate U-turn as scientist at centre of row admits: There has been no global warming since 1995.[152]

Why has this non-issue not simply dried up and died? Why isn't this article known to every English-speaking man, woman, and child on the planet?

Still disgusted...

Phil.: According to Planck a blackbody at 1500°F (~1100K) has a peak emission at ~2.7μm significantly in the NIR, whereas at 3000K it's 1μm so much more visible light, at 6000K it's 0.5μm in the visible spectrum. Hint, 1100K isn't 'hot'.

Mikael Cronholm: A 1500°F (815°C in real units) steel slab will radiate its peak emission at about 2.7μm (simplifying it as a blackbody, it's close enough). 2.7μm is in the infrared, not visual, just like Planck and Wien tells us. The people at Plant Engineering who came up with those numbers will have used the same laws (Stefan-Boltzmann, which is the zero-to-infinity integral of Planck) to calculate the numbers that you claim are falsified by the very same numbers. Really funny my friend!

It is the Sun that peaks in visual, at 480nm or thereabouts. Hotter than the Sun—even shorter peak emission.

And this:

> The Planck Law gives a distribution that peaks at a certain wavelength, the peak shifts to shorter wavelengths for higher temperatures, and the area under the curve grows rapidly with increasing temperature.

> If I've understood that correctly, if it's saying that the higher

---

[152] http://www.dailymail.co.uk/news/article-1250872/Climategate-U-turn-Astonishment-scientist-centre-global-warming-email-row-admits-data-organised.html#ixzz1EKbIpSQb

the temperature the more Visible light then it's been falsified already, see here:

The higher the temperature, the more of ALL wavelengths will be emitted. It is just the peak that shifts to shorter wavelengths.

Robert Clemenzi: To Mikael Cronholm, it is my understanding that the plastic in the desert condenses water on the bottom of the plastic, not the top. From *Collect Water in a Solar Still*:

> *Moisture from the soil then evaporates, rises and condenses on the underside of the plastic barrier above.*

Notice, this device increases the amount of water vapor under the plastic and does not collect water from the atmosphere.

I agree with you that a small amount of condensation occurs in the desert. However, in more moist climates, it is the morning dew and fog that release latent heat and keeps the surface warm—which explains why the dry deserts are colder. At any rate, my main point was that even "dry" air still has a lot of water vapor and George E. Smith was suggesting that "dry" means "none".

For Ken Coffman, for the emission at a given temperature, the gas spectra has to be multiplied by the blackbody spectra for each frequency. If the gas has a strong emission line at a short wavelength, but a blackbody would not emit at that wavelength, then neither will the gas. This is part of the reason why O_2 and N_2 don't play a part in spectral absorption and emission at 15°C, it simply is not hot enough. In fact, both O_2 and N_2 have absorption frequencies in the IR band, but their coefficients of emission are more than 8 powers of ten lower than the "active" gases. Those gaps in the spectra are not because no one has tried to measure the absorption there. Instead, it is because the absorption is so weak that our instruments cannot detect anything to measure. In addition, "normal" diatomic molecules do not have any peaks in the range we are discussing. The peaks that are there are because the two atoms in the molecule are different isotopes of the same element. For example, O_{16}-O_{18} will have a spectra in that region, but O_{16}-O_{16} will not. This spectra in the plots is low because the coefficients were adjusted to account for the natural abundance of O_{18}, which is very small.

Ken Coffman: Here is a bit of conversation about R. W. Wood's experiment—perhaps you've already seen it?

> *We may well ask if it is at all possible for backradiation to coexist as a significant process alongside kinetic transfer. It would certainly seem possible within the limitations of thermal gradients. However, if we revisit the experiment conducted by Robert Wood in 1909, an entirely different picture emerges. Wood constructed two miniature greenhouses identical in all but one respect. One used a plate of halite to transmit light into the interior, while the other used a plate of glass to transmit light into the interior (Wood, 1909). While glass absorbs more than 80% of infrared radiation above 2900nm, halite does not and is regarded as quite transparent to infrared. The point of the experiment was to test whether the halite's lack of absorption and re-emission of infrared radiation relative to that of glass would have any effect on the temperature of the greenhouse.*[153]

Mikael Cronholm: @ barn E, you make me homesick for my sauna back in Sweden! I am Swedish and my mother was from East Botnia in the Swedish speaking part of Finland, so sauna is in my blood so to speak. At my place in Sweden I built the two houses first and then the sauna, which is opposite to tradition—the sauna is built first in Finland! It is more than tradition, it borders on religion. The sauna is a home for spirits and the sauna gnome. The

---

[153] *The Shattered Greenhouse*, Timothy Casey, http://greenhouse.geologist-1011.net/

special sauna spirit has a name in Finnish, löyly, which is difficult to translate but it is the warmth and the feeling of well being, and it is surely connected with the steam too. It is an old word that traces back to Proto-Fenno-Ugric and has a counterpart in Hungarian meaning "soul".

By using evaporation and condensation the heat from the rocks can be transferred in a very controlled way. Sauna is not about sweating but condensation on the skin. The more water, the hotter. Löyly! Dry "sauna" has no löyly—so you are right. I have never seen one in Finland.

My sauna has neither electricity nor running water, even though it could be easily done. But it would take away the charm of the preparations, carrying the water in buckets, lighting the fire, the warm light from the paraffin lamp outside the window. You wash in the room itself, the water is mixed with hot from the heater on the chimney and cold from the buckets and it runs down through the floor and out on the ground.

Heating and cooling repeatedly is important. A snow bath is great!

Sauna makkara is sauna sausage, smoked, tasty. Beer and vodka can join it, but after or late in the session. You feel better then. Sit outside and cool down in the semi darkness of the summer nights in the north. Good sleep.

Here in Thailand where I live now it feels a bit redundant with sauna, unfortunately... (Sorry for going off topic!)

Domenic: to Robert Clemenzi, just two points...
1. I don't mean to be picky, I like your posts—actually dew is not just a morning effect, that's just when most people wake up and see it—it can go on all night. I live in the Miami Beach area, and I have often observed dew forming on my car immediately following sunset.
2. Regarding O_2 and N_2, you should have noticed that I also mentioned that they have to be tested for reflectivity and transmissivity at the long wavelengths. A gas molecule can be a very

effective 'greenhouse gas' even with little absorption—if its reflection to long wavelengths is high, reflecting them around to be absorbed by H_2O, etc, as well as reflecting them back towards Earth.

Don V: AJB says: *All your Joule are belong to us.*
I followed the link you provided to find an illustration comparing the energy content of water, CO_2 and N_2 vs. temperature, but unfortunately I have to disagree with the accuracy of what is depicted. The truth concerning water's amazing properties, that this illustration should show, is even better! Can you find all that is wrong with this illustration?

Mikael Cronholm: @ Ken. You got me all excited for a while there. "Halite", hmm, what is that? Never heard of it and Ken says its IR transparent! The "wow" factor was a little less when I looked it up—sodium chloride. So I learned a new name for it. Anyhow, very early IR lenses and optics were made from it. It is one of the few cheap IR transparent materials, but not the only salt. Barium flouride and calcium flouride are salts too, and IR transparent (at best only at 90% up to 12μm). But less cheap. Sodium chloride has the distinct disadvantages of being brittle and hygroscopic, so it cannot be used for making lenses for cameras. We are stuck with super expensive germanium, a metal, for the long wave cameras of today.

You could actually recreate that experiment quite easily by using polyethylene film—food wrap—instead of the halite. It should have similar enough characteristics to halite, at least compared to glass.

George E. Smith: Well as to my comments about "DRY" air; I am quite aware that real atmospheric air is never dry, in the sense that it contains NO H_2O molecules. BUT that is precisely what I mean when I say "DRY air". And that is simply to separate out the real King of greenhouse gases. CO_2 from the non GHG H_2O usurper, that is merely carrying the water (so to speak) for the true villain.

It also helps in that it removes the H_2O absorption spectrum from consideration when it comes to the heating and radiation from the atmosphere.

Somebody way up the forum there (my apologies to the poster; but you can find him) stated, that the atmosphere blocks (is opaque) in some bands in just the bottom few metres, and that that warmer lower layer radiates more heat downwards to the surface; whereas the higher cooler layers radiate much less upwards; ergo the atmospheric thermal radiation is concentrated downwards; and warms the surface.

So let's examining the validity of that notion.

Consider a very thin atmospheric layer anywhere in the atmosphere (doesn't matter where so long as it is in the lower 20 km or so); so for the moment shall we say our layer is at the standard Owl Box height of 2 metres; right where that thermistor is. A thermistor is a very poor excuse for a thermometer; but in this case it doesn't matter.

Shall we say our thermistor (refuse to call it a thermometer) reads 288K, 15 deg C or 59 deg F, the current assumed without proof mean temperature of the Earth. The S-B equation gives 390 W/m^2 (roughly) for this temperature, for the total Black Body Radiation. Did I say that my atmospheric layer is just one micron thick?

So it is emitting (isotropically) 390 W/m^2 upwards and 390 W/m^2 downwards (TIMES EPSILON, the EMISSIVITY). Now more accurately we should use the spectral emissivity; but for the time being we'll assume it is constant over the meaningful spectral range.

Now Wien's Constant (b) is 2.897756E-3 m.K +/- 8.4PPM, so that gives the peak of the LWIR spectrum to be 10.0616—I used to call that 10.1, I'm going to go with 10.0 microns.

Also the peak spectral radiance is given by $1.288E-11.T^5$ $w/m^2/micron$, which comes out to 25.51986. I'll take 25.52 Watts per m^2 per micron (of spectral bandwidth).

So now we are set to go. Taking the CO_2 band center as 15 micron (nobody knows what it really is); that makes it 1.5 times the peak (of 10.0 microns), and at 1.5 times the peak wavelength the BB spectral radiant emittance is 0.70 times the peak, which gives us 17.86 $W/m^2/micron$, at the CO_2 band. Getting generous, and giving CO_2 the entire 13.5 to 16.5 micron spectral width or 3.0 microns total, we get 53.6 W/m^2 (times epsilon) for the up and down halves of the isotropic thermal radiation from my one micron thick layer.

And that radiant energy is 98% contained between 5.0 microns and 80 microns (0.5 to 8.0 times the spectral peak wavelength).

Now that energy is radiated from the ordinary atmospheric gases of N_2, O_2, and Ar that are all at 288K temperature. And note that only 53.6 out of 390 is capturable by CO_2, that's 13.7% of the Total thermal energy that can be captured by the CO_2 in the layers immediately above and below my one micron layer; the rest goes right on through the adjacent one micron thick layers.

Now the layer above is colder than my layer, and less dense, so the Doppler broadening and the collision (pressure) broadening in that higher layer, are less than in my emitting layer (not much but less anyhow). The air layer below, is slightly higher temperature and slightly higher density, so both the pressure and Doppler broadening are higher than for my layer; and in particular are higher than for the layer above me.

So I have 390 (epsilon) w/m^2 of identical radiation going upwards, and downwards; but the layer above has narrower absorption lines than the layer below, so the layer below is going to undergo CO_2 absorption of slightly more of my 390, than will the layer above.

So both layers bordering mine, now absorb about 53.6 out of my 390 (don't forget the emissivity) but the upper layer absorbs slightly less, so that more of the upward LWIR energy goes beyond the upper layer, than goes beyond the lower layer.

Each of those layers will in turn thermalize that CO_2 captured energy, slightly raising their Temperature. Well not really; those thin layers will cool rapidly due to their thermal radiation losses, if it was not for the continuous resupply by the CO_2 GHG trapping process. That is why those layers are at those temperatures in the first place. Well don't forget that conduction and convection etc can also bring more juice to my thin layer.

Now I gave my layer some numbers; just to have some numbers to talk around; but it should be obvious, even if you have never scratched on the beach sands with a stick, that it doesn't matter where my layer is; so long as it is in the steadily declining density and temperature range of the atmosphere, the same situation applies. The escape route upwards, is favored over the return to Earth downwards, but those gradients and their line broadening effects. Okay, not by much to be sure; but certainly for sure.

It doesn't matter that the one micron layer right at the ground is radiating much more than the layer at 10 km altitude; it is still radiating isotropically, and the upper escape route is still favored over the return to Earth.

I'm NOT going to claim a measurable difference; it's the beat of a butterfly's wing thing; but it is quite real, and it puts the lie to the notion that the atmospheric thermal radiation is biased downwards towards the Earth; it isn't.

Now feel free to reintroduce H_2O; bring on your own favorite GHG; where's that fearsome 20 times methane monster.

I actually have a chart that purports to have a methane absorption spectrum. It has a very modest band at 8 microns, and a pitiful one at 3.3 microns, and a totally miserable one at 2.4 microns

Hey Earth to Atmosphere! There isn't any Earth emitted LWIR energy at 3.3 or shorter microns; and at 8 microns the spectral radiant emittance is 86% of the peak (10 microns).

Did I already say that only 25% of the total BB emission energy lies below the peak wavelength; and actually only 10% lies at less than .8; or 8 microns.

Yes in the hotter desert Temperatures, the thermal peak will move towards the CH_4 8 micron band.

You can do all of those things and though they change the absorption numbers; they do not change the fact that the upper cooler less dense layers intercept less energy than the lower denser, warmer layers; so the escape path is always favored.

And if you really want to believe that H_2O is a greenhouse gas, since there never is dry air. Whence comes this nonsense that H_2O needs CO_2 to kick it into action; and once your accept H_2O as a GHG all bets are off; because as the only condensing GHG, it is the only one that can form clouds, and then it can shut down incoming solar energy big time.

Our quite mild temperature range on Earth from -90°C to + 60°C, is regulated completely by the physical and chemical (bio too) properties of the H_2O molecule.(in all its phases)

@Robert Clemenzi. Robert, rest assured that I DO know that even in the driest of air, there is lots of H_2O. And thanks for your actual numbers for some interesting places. I simply wanted to make a "Dry Air", totally devoid of water, so that I could discuss the GH effects relating to just the CO_2 by itself; and I agree with you that the N_2, O_2 line spectra emissions are negligible in the scheme of things.

Once one understands how CO_2 in the $N_2/O_2/Ar$ atmosphere acts, then one can re-insert the H_2O or any other GHG and see that the same consequences still apply.

The point is that any small sample of air (one cubic micron for example), is constantly radiating a thermal spectrum; and by "thermal spectrum" I mean a black body like spectrum that is a consequence solely of the Temperature of the sample.

That radiation would of course cool the sample; but the energy is constantly being replenished by other energy input processes, either GHG absorption and thermalization or from latent heat or conduction.

So the air sample is at the temperature it is at, because that is the temperature at which its thermal radiation just equals the energy input from other sources.

In any case; thanks for your insights. If there was an app, for some geektoy or other that could compute the spectrum of any species in any environment to the extent that QED makes that possible, I would consider buying such a geektoy. Sadly I don't have access to some of the behind the pay wall programs; and also I have no confidence that they are even correct, since they never say what the circumstances of the derivation are.

Oliver Ramsay: It would appear that IR is selective in many ways. Apparently it's warming up my insides but it won't come out and tell me how warm they are. Either that or I'm about to succumb to hypothermia. My IR gun says my toes are 18°C. I'd always thought that was the surface temperature. I guess I really am cold-hearted.

Robert Clemenzi: @George E. Smith, we disagree on much of what you say. However, I do agree with "most" of your description of how more radiation goes toward space because the absorption lines get narrower. The problem with your description is that you ignore the clouds. Remember, they cover about 1/2 of the planet. Because clouds are blackbody radiators with no distinct spectral lines, there are no line widths to change with temperature and pressure. As a result, their downward radiation is greater than upward since the bottoms are warmer than the tops.

Also, please stop suggesting that O_2, N_2, and Ar are blackbody emitters. At the temperatures and frequencies of interest, they simply are not. On the other hand, clouds, aerosols, and dust are.

I am currently working on an app that works with the spectra of just water vapor and CO_2. If you would like to be an alpha tester, contact me via my web page.

Bill Illis: Robert Clemenzi says:

I am currently working on an app that works with the spectra

*of just water vapor and CO₂. If you would like to be an alpha
tester, contact me via my web page.*

I tried out the Stefan-Boltzmann convertor on your webpage.

One issue that should be relevant to this topic is that Earth is a rotating sphere.

During the height of the day at the equator, 1361 joules/m^2/second (less 30% Albedo) is coming in from the Sun but the surface temperature only increases as if 0.0017 joules/m^2/second is absorbed (or impacts the temperature at 2 metres). The extra 959.9983 joules/m^2/second flows away from the surface effectively almost as fast as the energy is coming in.

Your calculator says surface temperatures should increase to 87C.

At night, virtually no radiation is coming in (and the upwelling less downwelling radiation) says the surface should be losing about 100 joules/m^2/second but it actually only loses 0.001 joules/m^2/second.

This is the real-world now versus the theoretical.

The daily radiation budget fluctuates by huge numbers (±1060 watt/m^2) but the surface temperature only changes by ±5.0C on average or ±27 watts/m^2.

One SurfRad station measurements over a 24 hour period.[154]
[155]

Something else is going on in the real world.

Domenic: There is indeed, a LOT of that 'something else' that goes on in the real world.

That is why I have been repeatedly posting here that most assumptions made by climate experts, and others, regarding the 'greenhouse effect' and thermal radiation are completely wrong.

To Robert Clemenzi: For example, on your website you

---

[154] http://img140.imageshack.us/img140/4109/tablemountainall.png
[155] http://img12.imageshack.us/img12/3225/tablemountainnets.png

316

wrote: "On the Moon (no atmosphere and 28 day rotation), the maximum temperature is within a few degrees of what is predicted. However, on the dark side the minimum temperature is never reached—it is not even close. This indicates that objects heat up faster than they cool down. This asymmetry means that Stefan's law can be used to predict maximum temperatures, but is fairly worthless at computing the minimum temperature of a rotating body. It also means that using this equation to compute the expected temperature from the average energy will always give the wrong results."

The problem is not Stefan's law; the problem is the thermal mass of the target area and the thermal conduction within the target area material.

Here's an experiment for you: take a piece of Styrofoam and aim a radiant heat source at it while also aiming an IR thermometer at it. You will see the surface IR temperature immediately jump to an equilibrium value. Now, remove the radiant heat source, and you will see the IR temperature immediately DROP to the local radiational ambient temperature, virtually no time delay.

The reason, of course is that Styrofoam lacks thermal mass, so there is no thermal conduction to keep supplying the surface with additional heat once you remove the radiant source.

Stefan's law will hold perfectly in this situation.

Robert Clemenzi: Bill Illis says:

Something else is going on in the real world.

It is called the greenhouse effect. Conduction and convection keep the days cooler than expected. Radiation from the sky keeps the nights warmer than expected. Think of the poles, with 6 months of no Sun they should be close to absolute zero. It is heat radiation from a much warmer sky that keeps them at a "warm" 190K or so. In fact, at many places over land, the morning surface temperature is lower than the temperature of the air above it.

Mountain tops actually receive slightly more energy from the Sun than the land at sea level. Yet, they are colder because the air at that level is colder.

Where I live, in the spring it sometimes snows one week, will be in the 70F's the next week, and then snow yet another week later. During this period, the amount of energy from the Sun increases each day (remember, this is during the spring).
The difference between Stefan's equation and what is observed is the definition of the greenhouse effect.

AJB: Robert Clemenzi and Bill Illis say:

> *Something else is going on in the real world.*
>
> *It is called the greenhouse effect. Conduction and convection keep the days cooler than expected. Radiation from the sky keeps the nights warmer than expected. Think of the poles, with 6 months of no Sun they should be close to absolute zero.*

Certainly, it's clear that the relative warmth of the air over the pole doesn't come from sunshine at the pole when there isn't any. What's not so clear is that this story of radiating to the surface makes any sense at all.

It's the air temperature we're interested in. That's what we measure (or, guess at). That's what makes us put on our coats, or take them off. We are not running around sticking thermometers in the snow and declaring that the air has warmed it.

Conduction and convection are accepted mechanisms for cooling in the lower latitudes and that's how the arctic stays toasty.

The picture evoked is a static sky beaming down on the frigid landscape. The reality is more dynamic, even if the pace seems rather sedate.

The lapse rate is largely inverted at the pole, the radiative flow from the shorter atmosphere is upwards, the air circulates from equator to pole and back again.

The lower atmosphere at the pole is warmer than you might expect but not warmer than the air above it, which is radiating away to space, cooling and descending. Then it's slithering down south so it can get warmed up again. Sometimes, it passes through Tulsa.

The fact that some photons are absorbed by the snow is not very important.

Mikael Cronholm: @barn E. rubble, about your questions:

> For a start, the solar spectrum received at Earth contains 98% of its energy essentially in the wavelength range from 0.25 microns in the UV to 4.0 microns in the IR, with only 1% left over at each end. Ozone of course cuts off the short end at about 0.3 microns. About 2.5% of solar radiation lies below 300 nm.
> —George E. Smith

I totally agree with George on that, based on my understanding of Planck's Law.

I would agree that very little heat from infrared radiation would be absorbed by the surface of the Earth through the atmosphere, firstly because according to the above, very little energy reaches TO the atmosphere in the first place and secondly because even IF any significant IR reaches the "outside" of the atmosphere, it would have a real hard time getting through it. We have to keep the logarithmic scale of the Planck curves in mind when we look at them.

And then there is the so-called greenhouse effect that I believe is for real (although it has not been so clearly defined so I may be confused about it), and necessary for our survival on this planet. IR radiation from Earth will not radiate easily towards the cold space, but be trapped in the atmosphere to a degree. Some of the heat in the atmosphere must be in the form of latent heat in the water vapor itself and it keeps our planet from cooling off too much. This has been happening for millions of years, the balance has been shifting back and forth and warmer and colder periods have been taking place for a variety of reasons—in combination! I am just not very sure about these mechanisms, but somehow there is a balance between heating and cooling that keeps us at a comfortable equilibrium.

I think it is important to make a distinction between "green house effect" and "global warming". Greenhouse effect as such is necessary and not an evil thing, if the term means what I think it does. We depend on it. The other claim that CO_2 emissions cause an additional warming effect—that is compounded with an increase of the greenhouse effect—is the part that I feel is not proven at all. This is where all the politics and prestige comes into the picture and as a crowning of it, even fraudulent behavior from scientists.

So there we go, I think that in addition to replying to your very well chosen questions I was also prompted to sum up my personal standpoint at this time. Subject to continuous scrutiny and revision of course!

Phil.: Design an experiment that simulates the Greenhouse Effect and we'll have something to talk about, the experiments you describe are far removed from that. You'll need to have a light source corresponding to a blackbody at 300K (peak wavelength ~10μm), and a container transparent up to ~20μm (ZnSe would probably work although a ½" diam window will cost over $100). Have at it.

Mikael Cronholm: @ Phil. Creating a 300K source is no problem at all, since it corresponds to about 27°C. The problem is to simulate the atmosphere and space. That is not something you do in a kitchen!

In the end, the burden of supplying proof rests instead on those that claim CO_2 is responsible for warming the Earth. None of that has been presented so far and it will be difficult to do. It brings me back to what I ended the discussion with Ken with:

> *...the problem that all these scientists have is that they will never be able to test if their theories are correct, because the time spans are too long. For a theory to be scientifically proven, it has to be stipulated and tested, and the test must be repeatable and give the same results in successive tests for the theory to be proven.*
>
> *If not, it is not science, it is guessing.*

Robert Clemenzi:
www.coe.ou.edu/sserg/web/Results/Spectrum/o2.pdf and
www.coe.ou.edu/sserg/web/Results/Spectrum/n2.pdf show the IR
spectra of O_2 and N_2, respectively. Unfortunately, there are no
spectra for H_2O and CO_2 for comparison.

Actually there are at the same site that you get those from. To save
trouble here are the spectra for CO_2, N_2 and O_2, plotted on the
same graph[156] and here is the blackbody emission spectrum on
approximately the same scale:
http://i302.photobucket.com/albums/nn107/Sprintstar400/BB.p
ng
     Note that the scale of the first graph is log base 10 so that the
N_2 and O_2 bands are many orders of magnitude weaker than the
CO_2 bands, also in the region where they show up is an H_2O peak
which also swamps them by a similar margin.

Mikael Cronholm: A white car and a black car will have the same
absorptivity in the IR, and yet the black one gets hotter when they
stand in the Sun beside each other. The white car reflects more of
the visible light from the Sun, while the black one absorbs more of
it. Is that impossible for you to understand?[157]
     Well, there is a set of three curves (solar radiation on the

---

156

http://i302.photobucket.com/albums/nn107/Sprintstar400/CO2N2O
2.png

[157] http://en.wikipedia.org/wiki/File:Solar_Spectrum.png

outside of atmosphere and on Earth, compared with blackbody) that will show you that the Sunlight has the most intense energy flow (which becomes heat when absorbed by the Earth) in the visible part. There is also a lot in the near IR, but not much over 2.5μm, which is what I pointed at regarding Herschel. There would have been very little energy in the longer wavelengths and they would have been blocked by the glass prism anyway.

BTW: "The Planck "peak" might well be in the shorter wavelengths." It may, it may not—it depends on the temperature of the object where the peak ends up.

Phil.: Here's the absorption spectrum for liquid water (blue curve):
http://www.btinternet.com/~martin.chaplin/images/watopt.gif

Note the 8 order of magnitude drop-off from the near IR to the UV (300nm).

Mikael Cronholm: @ Phil. Thanks! I had a feeling it should look something like that.

Domenic: to Mikael: This paper might be interesting to you.
Infrared and Sub-millimetre Observing Conditions on the Antarctic Plateau[158]

> *The temperature of the atmosphere affects the flux levels in the near IR, mid IR and sub mm IR quite differently.*

Don V: AJB says:

> *Apart from conflating the impact of the illustration with sublimation and pressure change, no. The data largely came from Engineer's Toolbox (e.g. for water vapour). Feel free to elaborate and correct, I'm all ears.*

---

[158] http://www.phys.unsw.edu.au/jacara/Papers/pdf/pasa_modtran.pdf

Okay, the Y axis is in units of heat capacity. The heat capacity for ice is illustrated correctly, the heat capacity for water is illustrated correctly, the heat capacity for steam is illustrated correctly, and the heat capacity for N_2 is illustrated correctly all at 1 ATM. But, at 273 K the latent heat of fusion shouldn't be illustrated with just a big orange blob. In fact at its melting point ice just sits there and sucks up 334 kJ/Kg with no change in temperature at all. This should be illustrated on a logarithmic Y-axis scale showing a huge spike in energy absorbing capacity—a vertical line that jumps not one but two decades from the 2.0 value up to 334 and then back down to eventually arrive at the 4.2 value of water.

And then again at 373 K it should show an even bigger spike of 3 decades of energy absorbing capacity of 2257 kJ/Kg to get from water to steam.

In addition, CO_2 is incorrectly illustrated at 195 K. This is the temp at which it sublimes. Below 195 the heat capacity of solid CO_2 is 1.2 kJ/(Kg*K). Right at 195 a similar spike in the graph should be illustrated of ~760 kJ/Kg to reflect the latent heats of melting + vaporization since CO_2 sublimes at 1 atm.

But neither of these adequately illustrate what is going on, because the units of latent heat of fusion, and latent heat of vaporization, don't match the units of heat capacity.

I believe what the author of this illustration was trying to show was how much energy absorbing capacity water had over either of the atmospheric gases. IMHO to illustrate this correctly, the Y axis should have just had units of energy, and the X axis units of temperature. When this is laid out properly, even with a logarithmic Y axis, the significantly higher heat capacities of ice, water and water vapor, PLUS the huge energy absorbing phase transitions, show just how amazing water's temperature/energy buffering capacity is! You need a logarithmic scale just to keep the other gases even on the graph, otherwise they look like they are both barely above zero over the whole range of temperatures. (BTW: the normal climatic temp range is 240-320K)

I would like to add one more comment to this thread that suggests a neat experiment anyone can do to prove the very low contribution of CO_2 back radiation to our blue planet.

http://solarcooking.org/plans/funnel.htm

On this page (put together by BYU professor Steven E. Jones, and his students) you will find simple directions for how to build a simple solar cooker out of a piece of cardboard, aluminum foil, a mason canning jar and a "shake and bake" bag. The plans illustrate a way to make a simple funnel shaped mirror with the canning jar (blackened) sitting inside the bag (to prevent convective energy loss) at the focal point of the mirror to focus radiant energy transfer and collect all of the energy in the Sun's light rays.

Now, you could play around with this fun toy on a nice sunny day (or even a cloudy day) and show just how much radiant energy rains down on us. But that isn't the cool experiment. No, the cool experiment is to take the funnel mirror with the blackened Mason jar (a blackened plastic frozen juice can might work better) inside of not one but two plastic bags out on an average cloudless night, and face the mirror out at the black sky.

Read almost to the bottom of the page at the above web site and you will see that over a couple of hours the radiant energy transfer to the cold black body temps of deep space will drop the temperature of the water in the can 10 degrees C below ambient! They claim they even achieved freezing!

AJB: Hi Don, point taken about CO_2 at 195K. But you seem to be having a problem with units, the blobs as labeled bear no relation to the Y-axis and temp goes nowhere. I guess you'd rather see something like this (no need for a log scale, did you forget to factor out K on the Y-axis?).[159]

Obviously both depictions are hugely simplistic, just a like for like comparison of each in a pure state, hence the jokey title. Water is nifty stuff though, except when you don't want it playing it's big

---

[159] http://img580.imageshack.us/img580/2442/water2.png

bag of tricks :-)

Domenic: To AJB: Nice graph. Now graph the same three molecules with respect to number of molecules of each gas within the atmosphere—for total joules in the atmosphere for each.

What do you see?

Focusing on CO_2 by the warmistas looks pretty ridiculous once you look at the whole scheme of things....

Ken Coffman: AJB, I second Domenic. We are very fortunate to live on a planet with so much water—which is capable of absorbing and emitting huge amounts of energy during the two common phase changes—all of which help us maintain a livable temperature. Energy does not equal temperature. The reason we're here today thinking about this stuff is a side-effect of that fact.

Don V: AJB, yes, that's closer to what I'm talking about! Are you sure you got all the slopes and phase change jumps right though? Especially for CO_2? Remember the energy content can be computed as $m*Cp*(T-Tzero)$ up to the first phase change. Thereafter the energy content is equal to the total heat capacity at the temperature of the phase change plus the phase change energy plus the heat capacity of the new phase multiplied by the delta T as measured from the phase change. The CO_2 graph should include a large phase change energy added to the solid phase heat capacity starting at its sublimation temp.

Furthermore water's slope should be 4, ice to water jump only 334 (graph seems a bit more than that), and water to steam jump 2257 (graph should hit near 3600 near 400K).

Domenic, I agree, that's the point I was trying to make. Multiplying each of the numbers in each graph by the mass of each gas in a kg of air would illustrate the energy content carried by each. Since CO_2 is such a tiny, tiny fraction of air's total volume, its energy carrying capacity is miniscule, when compared to water's. The difficulty of producing this kind of graph though is deciding what part of the atmosphere you want to "sample" and deciding on the sample's water content. It can go anywhere from near zero in the dead of winter at the poles to greater than 1 Kg(H_2O)/Kg(Air) in a drenching ice storm (all three phases).[160]

And then when you add in the density differences of each of them in the gas phase and realize that most of the CO_2 is sinks to the bottom.

Domenic: to Don V, we do have a pristine data baseline reference to gauge the true 'greenhouse effect' of CO_2. It's the Antarctic data. (Note: do not use Antarctic coastal data, it's too noisy due to higher average temps and higher humidity. The interior data is most pristine. High signal, no noise.)

There is virtually no moisture in the atmosphere above Vostok and Amundsen-Scott.

And there has been a rise in CO_2 in the air there comparable to Mauna Loa.

So, in Antarctica, any 'greenhouse effect' of CO_2 increase should be magnified there, as there is no water vapor to moderate it.[161]

Keep in mind, the 'greenhouse effect' of CO_2 the warmists keep pointing at actually has two components (1) absorbing more incoming solar radiation during daytime, and (2) keeping heat 'trapped' during nighttime.

And yet, the temperature records there show decreasing temperatures, not increasing with increased CO_2 also present there.

And during both the Antarctic summer and winter! Sun and no Sun.

What bothers me most is that Keeling must have noticed that. He was involved in setting up the original CO_2 monitoring in Amundsen-Scott in 1957. That SOB must have known his original

---

[160] http://www.engineeringtoolbox.com/moisture-holding-capacity-air-d_281.html

[161] http://icecap.us/images/uploads/VOSTOK.pdf

theory of connecting global warming with CO_2 was a sham in his later years, prior to his death. And yet, he said nothing...

barn E. rubble: And here I thought everyone had moved on from this thread. (I got real busy and didn't get a chance to catch up until now.) I appreciate the time and efforts from all who've contributed here. I'm sure I'm not alone and there are many more who think the same, reading thru all of this with great interest. As much as it may not seem to those directly involved but—this is real debate! Science vs. science (in spite of how some may see it) arguing about how things 'REALLY WORK'. Well, from their point of view— which of course is ALL debate. I picked up some animosity but no incivility. I think those involved in the details here may underestimate how many people were following your debate and just because they didn't post a comment doesn't mean you were arguing between yourselves. This has been a great thread! It has been enlightening and disheartening, i.e., for those of us looking (hoping?) for agreement that 'all is well and here's why' vs. the 'we're all gonna die and it's my fault.'

While trying to keep up with those here RE: graphs, laws and such, I tried to find some base understanding of the topics discussed by reading up on the specific stuff mentioned here, elsewhere. Thinking the more sources of information I found the better. As if! It seems the more I read the worse the fog gets!

On my recent journeys I found many sites that left more questions than answers. Unfortunately (for me anyway) well written pieces can seem quite logical and reasonable. Thanks to the internet, i.e., post whatever ya' want, you need to have your own info/bullshit filters. Only you can't buy them. So those like me are left to try and make our own. One area in particular that I've read about seemed reasonable but because I hadn't seen/heard about it before leads me to think I may have been misled. As in my own BS meter started to twitch.

I'm referring to the Stefan-Boltzmann law which is something that has been a major foundation block (if you will) to the discussion here. Now this may well be old news to most here but it's the first I've read about NASA finding the SBL of no use for the moon landings.

For those that know better, is this just a 'junk science' thing that I've read? I found a number of sites/pages starting from about a year ago and they all read well—seem to have supporting references and 'non-crazy' authors. I have no way (either education or pro experience) to refute what I'm reading.

What are the thoughts here on the premise that SBL has no 'real' world application? And in particular how real world applications of SBL on the AGW argument?

I've gone back thru this whole thread looking for this connection and didn't find it. Maybe I missed it.

Again to those contributing here; I'm not the only one who appreciates your time and expertise here.

Mikael Cronholm: @ barn E. rubble and other readers/contributors. I totally agree with you, it has been a very interesting debate here, and I second barn E. rubble in the appreciation of all the input.

The laws of physics are not depending on us for their existence. We don't make them up, we discover them. They are not inventions, they are the rules and mechanisms by which the universe operates. Once discovered and found to hold true by testing, it is up to us to use them within their proper realm and limitations. So a law of physics does not care whether it is "applicable" to us or not in a given situation. We have to figure that out, and to do it, we first need to understand it. Really understand it. I have realized just how big a challenge that is for people who are not used to working with these laws regularly for years on end. I see the same challenge towering before the students I teach in IR thermography. The difference is that we have material prepared, we have the time we need to go through things step by step, so that all is introduced in a less painful manner and gets into place.

Trying to get things explained well on a forum like this is

difficult. There is a great variation of previous knowledge between readers and you cannot find a balance to suit everyone's needs when explaining something. It will be too simple or too difficult for many readers. Of course, the big plus with a discussion like this is the interactivity which weighs up some of the disadvantages.

In general, I have a feeling that a lot of the stuff that is written on these topics are going over our heads. Scientists in the higher hemispheres are not very good at explaining in layman terms and that presents a challenge to all of us.

The saddest thing though is when science becomes hostage of politics, and politics are driving science instead of the opposite. Scientific discovery cannot be planned, ordered, administrated, or made to fit preconceived models. That is the problem with the global warming issue. Scientific work needs to start with a hypothesis, of course, but the result of the research must be accepted whether it turns out to prove the hypothesis or not.
I wish everyone best of luck in their endeavors of learning!

Robert: Hello. Progress of thermodynamics has been stimulated by the findings of a variety of fields of science and technology. The principles of thermodynamics are so general that the application is widespread to such fields as solid state physics, chemistry, biology, astronomical science, materials science, and chemical engineering. I found two great open access books: *Application of Thermodynamics to Biological and Materials Science* and *Thermodynamics*. You can read them on an online reading platform or just download them.[162] [163]

The contents of these books should be of help to many scientists and engineers.

ThomasU: I too want to thank all participants for this great discussion!

I certainly experience the difficulty described by Mikael

---

[162] http://www.intechopen.com/books/show/title/application-of-thermodynamics-to-biological-and-materials-science
[163] http://www.intechopen.com/books/show/title/thermodynamics

Cronholm also by barn E. rubble and perhaps many others here (I didn't have the time to re-read the whole thread)—the difficulty to follow a discussion which partly is well above my head. The one thing which is obvious nevertheless, is that "the science is settled" can only apply to science settled to follow the scientific principles. Any appeal to authority, any smearing of differing opinions, any "untidiness" resulting in lost data, any group-thinking, "save-the-world" fervor, etc.—should be ruled out and disqualifies the person who resorts to it. For me at least.

Ken Coffman

Uncommon Common Sense

*How should we deal with flaws inside the climate community?
I think, that "our" reaction on the errors found in Mike Mann's
work were not especially honest.*
—Douglas Maraun, Associate Fellow, University of
East Anglia, Climatic Research Unit

Interview with Geologist Don Easterbrook

ON MARCH 27, 2008, in Burlington, Washington I
interviewed Don Easterbrook.[164]

Don J. Easterbrook: Some people say that global warming
skeptics think the moon shot was staged and the Earth is flat...
Ken L. Coffman: Funny you should mention that, here it is, I have
the exact quote.

> *Al Gore: You're talking about Dick Cheney. I think that those
> people are in such a tiny, tiny minority now with their point
> of view, they're almost like the ones who still believe that the*

---

[164] Dr. Donald J. Easterbrook, Professor Emeritus Geology, Western
Washington University, author of 8 books, 150 journal publications with
focus on geomorphology; glacial geology; Pleistocene geochronology;
environmental and engineering geology.

331

*moon landing was staged in a movie lot in Arizona and those
who believe the world is flat. ... That demeans them a little
bit, but it's not that far off.*
—*CBS-TV, 60 Minutes, March 30, 2008*

KLC: I was going to ask for your opinion on it…
DJE: Online, you will find ten talking points about what Gore has
said and it essentially points out that what he's saying is a bunch of
hogwash. It's been refuted by the scientists who work in such
things.
KLC: I wanted to talk to you about Al Gore because you seem, in
general, to have been supportive of him.
DJE: Actually, I'm not. The irony is that I voted for him [in 2000];
I'm neither a Democrat or a Republican. I dislike the Democrats
only slightly less than I dislike the Republicans, so I'm one of those
independents who think the government is totally corrupt in both
parties. The point being, simply, I don't have a political agenda one
way or the other.
KLC: In some of the stuff I've read, you seemed to be defending
him. Like you said, you voted for him…
DJE: Actually, it's not that at all. For a number of interviews,
especially in the national news media, they ask 'Are you a
Republican?' and I say 'No, I'm not, as a matter of fact, I voted for
Al Gore. I don't want to pick on him because he's not a scientist.'
The crap he spews comes from the IPCC[165], so in a sense I don't
condemn him as much as I do the so-called climatologists like
[James] Hansen, he's an idiot. They're the ones giving him all this
stuff. He's a propagandist, not a scientist, so I cut him a little slack.
The things he does, the things he says are so outrageous, I don't
forgive him anymore because, when he says things like 'people like
me are right in there with the flat Earth theory', to hell with that.
He says the debate is over. The debate is not over, there's a huge
uproar in the scientific world because in the last ten years, the

---

[165] United Nations Intergovernmental Panel on Climate Change

climate has cooled and the media won't tell you that. This year is a big downturn, you can't miss it. Global warming simply ended in 1998, but the public doesn't know it.

KLC: I could draw it myself, you have a peak in '98 and it's been flat or declining since then. The trend depends on where you start. They love starting in 1850...

DJE: That doesn't work because there are 30-year cycles. The chairman of the IPCC admits we've had global cooling for at least eight years, and there are sources on the Internet, you've probably seen them, that show the IPCC folks are panicking.

KLC: Talk about an inconvenient fact...

DJE: On the temperature curve, 1998 was the high point, and by 2006 we've cooled dramatically for a couple of years. It's been kind of flat. A sort of a plateau, but if you take 1998 as your starting point, it's down...

KLC: Playing the devil's advocate, if you start in the early 90's, you would still have a positive trend.

DJE: If you want to be really honest about this, the curve should rise from 1945 and the bottom should fall out after 1998. It depends on what you want to show and how you want to filter it. You can filter it with a two-year average, a five-year average, or over whatever period you want and you'll get a differently shaped curve. The point is, it has not gotten warmer since 1998; it has not continued to warm in the last ten years.

KLC: How can that be if CO_2 is increasing?

DJE: You can take ground data or satellite data; they are not exactly the same, but close. I took all the data, satellite and ground, averaged it, and plotted a single curve from the average to show the trend and the bottom falls out in 1998.

Look at history, where we are in 2008 and where we've been. If you go back to the beginning of the century, there was a really deep cold period from about 1880 to 1910, and then it warmed to about 1945. Most of the global temperature records are set in the middle of the 1930's when it was warmer than now. And the same is true in Greenland, the temperatures in the 1930's were warmer

333

than they are now. In 1977, we did a flip to thirty years of global cooling. The time of maximum CO_2 emissions started in 1945 and temperatures should be shooting up, but we cooled off. That's an anti-correlation. In 1977, we get warmer and warmer and warmer. If we look back 500 years, the trend of 1977 to 1998 is not unique to this century, for hundreds of years we have 30-year periods where it gets warm/cold, warm/cold.

We've been coming up about a degree per century since the Little Ice Age in about 1600. We've been warming for 400 years, long before human-generated CO_2 could have anything to do with the climate. If we project the previous century into the coming one, my projection, from 1998, is that we have about a half-a-degree of cooling from 2007 plus or minus three to five years to about 2040. Then it starts getting warm as we enter the next cycle, followed by cooling again. By the end of the century, we'll have less than a half a degree increase, instead of ten degrees. A huge difference. The IPCC projection says we should be one-degree warmer than where we were in 2005. But, we're getting colder. We declined almost one-degree in one year. We're going in the opposite direction. With IPCC data and their graph, by 2011, the difference between my projection and theirs is about one-degree and that's huge. Now, they have to increase a degree in three years. If that doesn't happen, they're projection wrong and mine is right. By 2038, the difference is two degrees.

KLC: Around 2002 you predicted cooling. That must have been a tough thing to come out in public and say in those times.

DJE: I was a lone wolf howling in the wind in 1998. I gave a paper in 2001 in Boston at the national GSA [Geological Society of America] meeting and you should have seen the stunned look on people's faces. We'd just had 1998 warmth peak and people were astonished. I said, look at the data and forget CO_2. You know how much change there's been in atmospheric CO_2 since the advent of big man-made emissions?

KLC: Maybe 100PPM [Parts Per Million]…

DJE: Not even that. Normally it's been about 280PPM. It crept up to about 300 by 1945, which is nothing, it had been naturally that high before, and in 1945 it took off. Emissions went straight up. However, the total change was not much compared to the volume of CO_2 already in the atmosphere. Water vapor is the main greenhouse gas and one of the things you won't hear anywhere is that in order to get the global warming projected, the CO_2 people can't get there with only CO_2, the effect isn't big enough, so they say it will change the water vapor. They rely on water vapor to get their climate change, not CO_2, CO_2 is just enough to nudge it and water vapor does the rest.

KLC: At Real Climate[166], I've said that the idea of CO_2 being a 'forcing' and water vapor being a 'feedback' is great marketing, but bad science. You can imagine what response you get from something like that. I think it's brilliant marketing.

DJE: Look at the difference, man contributed eight one-thousandths of one percent to the total CO_2 that was present before the big upshoot. It is instructive to think about emissions, not atmospheric content. From 1870 to 1900, we had global cooling, then we had significant global warming from about 1910 to 1945. This global warming is not accompanied by any significant rise in CO_2, so you can't blame it. Then CO_2 increases while we have global cooling. You can't blame *that* on CO_2. It's only been the last 30 years there's been correlation between CO_2 and global warming. Everything before was uncorrelated. There's no doubt there's been warming as we come out of the cold period from 1880 and 1910. The 1930's were warm, then we cooled from 1945. 1977 was a turnaround year when temperatures started up and now we're headed down again. We're right where we ought to be.

As an aside, you know the big news from the Antarctic about the Wilkins Ice Shelf breaking up? The headlines in some news media was something like 'Antarctic Ice Sheet Collapses'. It's not the ice sheet which is 15,000 feet thick. What we're talking about is really thin shelf ice along the margin which warmed by ocean

---

water. We've had thirty years of global warming, the water is getting warmer. So what? The truth is, the main Antarctic ice sheet is getting colder. The snow records show the same thing. The ice is not shrinking, it's growing. Al Gore says it's 40 degrees down there and everything is going to melt and the sea level will rise. Hansen has said the sea level could rise as much as 70 feet. It's insane. Absurd.

KLC: They love talking about the area that sticks out into the Pacific-Atlantic intersection.

DJE: The shelf ice is really thin. In terms of total volume, it's nothing, a fraction of one-percent. It doesn't mean anything. It's what you'd expect if you have warm ocean water. It's impressive because it has a broad area, but it is really thin. The climate is not warmer down there, the surrounding ocean is warmer. Al Gore was quoted as saying something about talking to somebody who said it was thawing in Antarctica where it was something like 40 degrees. Someone bothered to look up the temperature. At that point on the ice cap it was 47 below zero. The caller was down on the coast someplace by the nice warm ocean water.

Look at Greenland. Both Gore and Hansen talk about glaciers sliding into the sea. That's crap, no glaciologist in the world would subscribe to that nonsense. Glaciers don't do that. Two-mile thick ice, almost three-miles thick in Antarctica, flows like really thick taffy, and it doesn't slide anywhere. It's like saying Pike's Peak is going to slide into the Gulf of Mexico, it isn't going to happen.

KLC: They can't talk about the Arctic, they have to talk about ice on land, Greenland for example.

DJE: There are no glaciers in the Arctic. There's a big noise that Gore and Hansen made about melting in Greenland. There is melting along the edges, but the ice is growing in the middle, like Antarctica. To discuss Greenland temperatures, we had global warming from the turn of the century to the 1940's, then Greenland experienced the same global cooling that everyone else did from 1945 to about 1977 and it's been warming since then. The interesting thing is, it was warmer in Greenland in the 1930's than it is right now. They're saying it's never happened before? It happened in the 1930's.

KLC: Speaking of Greenland, I'm curious about your take as a geologist about what you see as a driving factor. From my reading I know you believe there is a correlation with solar cycles, but I'm curious about the geothermal heating.

DJE: The temperature is too variable to be accounted for by volcanic activity. Aerosols from eruptions last two years, and then they're done. With regard to geothermal heating, there is geothermal activity in Greenland that may be contributing to the melt, but I suspect the root cause of melting around the edges is surface temperature. We're in a global warming period, what do you expect to happen?

I plotted a curve from isotopes in Greenland. Oxygen isotope ratios give us ambient air temperature in the snowfall. The isotope signature doesn't change, so you can core ice and examine annual dust layers that mark the ablation surface in the summer when the dust settles on it. You can identify the melt season very accurately and go back for thousands of years with one or two year accuracy. The chronology is very accurate. We don't have it in the Antarctic, but we have it in Greenland. I calculated the Oxygen 18 ratios and I show times when the temperature was warmer and colder. I marked the bottom of every cool period. If you calculate the cool periods, they are about 30 years apart. Which means the same thing we saw in the last century has been going on for 500 years. It's nothing new.

Let's discuss today versus the past. During the medieval warm period [MWP], about 900AD to about 1300AD, climates were warmer than they are now. One thing Gore talks about is the so-called hockey stick, a curve made by [Michael] Mann, a tree ring specialist. He essentially shows flat temperatures until we get to the increase in CO_2 where the temperature takes off so it looks like a hockey stick. His data was examined by a panel appointed by the National Academy of Sciences and found to be totally spurious; they said it is not scientifically sound and threw it out. The reality is, temperatures go up and down in a regular pattern; it isn't flat.

In a book by Fagan[167], he wrote about the Little Ice Age and the Medieval Warm Period. He confirms much of the evidence about glaciers that were way down-valley and way up-valley in the Medieval Warm Period, which was warmer than now. Then the temperature plummeted about four degrees in about 20 years. Boom! They went from a time when people in Europe were thriving; for example, there were colonies in Greenland and they made a lot of wine in England and shipped it to France. Suddenly, we're in this Little Ice Age and a third of the population perished in Europe, not all from starvation, there was plague and a lot of other things that was made worse by the starvation. There was incredible famine, hundreds of thousands of people actually starved to death.

We were warm from 900 to about 1300 where we start the Little Ice Age. In 1609, Galileo perfected the telescope and could see the sunspots for the first time and scientists began recording sunspot numbers. The number was very small. There is a direct correlation between sunspots and solar irradiance, the energy we get from the Sun.

Astronomers predict the start of Solar cycle 24 and they keep shifting the curve because the it's not happening. Normally you have more than a hundred sunspots per year and we're near zero right now. The current sunspots are from Cycle 23. Cycle 24 was supposed to begin in March, then they pushed it back to May and some people are saying they'll push it back until September.

KLC: What do you think, Don?

DJE: I don't know and there is no way to predict it. Look what's happening to sunspots and to temperature. I show correlation sunspots and global temperatures.

How do you explain increasing atmospheric CO_2 when we had global cooling from 1950 to 1977? Prior, you go from cooling to warming without any change in CO_2 at all.

Here's the answer to the hockey stick: temperatures from the

---

[167] Brian M. Fagan, *The Little Ice Age: How Climate Made History 1300-1850*

Greenland ice cores take us back about 15,000 years and show our current period, the Little Ice Age and the Medieval Warm Period. We have increases of up to 23 degrees in a century and the same amount of cooling, and another increase of 20 degrees of warming in a century. The idea we've never seen changes in global temperature like recent changes is totally fraudulent. Gore still says it. There's a big dip 8,200 years ago showing about 2 per mil Oxygen 18 which is equivalent to a couple of degrees, nothing like the ice ages. There was a cold period that peaked in 1890. The isotopes follow the recorded temperature, which is a check on how accurate the isotope readings are.

Glaciers are wonderful for recording climatic fluctuations. In 1609 when Galileo perfected the telescope and the record shows, generally, 50 to 100 sunspots. We see the Maunder Minimum when virtually no sunspots were recorded. We see the global cool period from 1945 to 1977, the warm period of the last thirty years, the 1890 cold period. Sunspot activity mirrors global temperatures almost exactly. The curves dance together. I can't imagine anything tighter.

KLC: What about the claim by Hansen and others that TSI [Total Solar Irradiance] has been unchanged for 80 years, thus it cannot explain recent global warming?

DJE: Not true. If you look at the data coming out, you see a strong correlation between global temperature and irradiance. If you plot irradiance versus sunspots, you again get the same kind of curve, and the inference is you can connect these curves to recognize a link between global temperatures and the sunspot cycle. We only have satellite measurements back to about 1970, thirty-some years of data, the change is about a tenth of one-percent. That's more than the eight one-thousandths of a percent of CO_2 change, so they say the TSI change is not enough?

The argument I make is the correlation between sunspot activity and temperature is not fortuitous, it can't be. There is a cause and effect relationship. We don't know what the connection is, but it is obvious that a small change in solar irradiance produces a big climate change. It's leveraged by something, maybe by water vapor. We're not sure. The argument that it's not big enough? A friend of mine has a saying which I love. "If it happened, then it must be possible." Well, it happened, so it must be possible.

KLC: The AGW crowd uses the same argument, CO_2 rose and temperatures increased. It happened.

DJE: Only in the last 30 years, before that you have the opposite of correlation. There is correspondence between solar activity and global climate. We don't understand the connection and the leverage and it doesn't seem like it should be enough, but it *is* enough. The Little Ice Age overlaps the Maunder Minimum solar irradiance, not measured directly because we don't have sunspot numbers from back then, but when you have a change in solar irradiance, you change the amount of Beryllium-10 produced in the upper atmosphere and radiocarbon, so you can use those measurements in ice cores as a proxy for with solar irradiance and sunspot cycles.

Correlation over the known period of observation establishes the linkage. Fluctuation in radiocarbon and Beryllium-10 from the upper atmosphere shows the change in the amount of solar radiation. The Dalton Minimum occurred in about 1830. 1890 had a TSI minimum and then here's 1945 to 1977, the correlation is there too. No one argues that the LIA [Little Ice Age] was not caused by a change in solar irradiance represented by the Maunder Minimum. If that's certain, then why not the others, including the current warming period? There is a strong case for solar control. There is a symposium in Oslo in August where a bunch of us will present papers spelling out the relationship between global climate change and solar changes. There are a growing number of scientists, especially astrophysicists, who are convinced the climate driver is solar, not CO_2.

Here are the IPCC predictions, up to ten degrees hotter by the end of the century. Here is mine, a rise of about half a degree. Let's place a checkpoint at 2011. They need to see another degree of increase, or they're wrong.

KLC: They'll move the prediction around...

DJE: Every year they recalibrate their computer model and put in the observed temperature. So, as they go along, the curve that trails behind is perfect. It's like predicting the morning's weather at six-o'clock in the evening.

KLC: They call it hind-casting. It's clever. Use the same model, but reset the starting point each year.

DJE: They published their projection, so I'll hold them to it.

KLC: They're slippery. I look at it from an engineering standpoint and so much of it seems absurd. I don't understand how they get away with it. Mass psychology and herd instinct?

DJE: Do you know what drives them? Money.

KLC: You're talking about research grant money?

DJE: I'm talking about money, period. Al Gore, before he started all this, was a millionaire, worth two or three million. Now he is worth 100 million dollars. He has a slush fund of about five-billion dollars his hordes have gathered. He sends out people as emissaries with his slides so he doesn't have to. He gives few interviews, though he's appearing on CBS 60 Minutes. If you're in the press and you want to attend one of his lectures, you can't. Not only can you not ask questions, they won't let you in. Because the debate is over, you see. You'd just be a troublemaker. You know about the Bali-100?[168]

KLC: Yes.

DJE: You know about the 400 consensus breakers?[169]

KLC: Yes.

DJE: There's a new one called the Manhattan Declaration.[170]

---

[168] A letter disputing the findings of the IPCC sent to the UN Secretary-General and signed by 100 scientists.

[169] 400 scientists voicing objections to so-called "consensus" on human-caused global warming. These scientists are listed in a report issued by Sen. James Inhofe, R-Okla, who is on the U.S. Senate Environment and Public Works Committee

[170] The Manhattan Declaration comes from the 2008 International Conference on Climate Change and suggests world leaders reject views

KLC: Don, you know they're all in the pocket of big-oil. They believe in a flat Earth.

DJE: I get so much Bush oil money that I'm embarrassed to go to the bank. I push a wheelbarrow.

KLC: I note you're carrying a Conoco-Phillips bag...

DJE: So I am. It's from a GSA meeting. It also has Halliburton on it. For the record, I've never taken a nickel from any industry. It's the first thing interviewers ask. I'm making a lot of new friends to the right of political center because they love what I have to say. The thing they love most is that I've never taken a nickel from industry. I'm not a Republican, so I must be pure of heart.

KLC: It's too bad it has to be so political. I'm just interested in the science.

DJE: Al Gore makes a hundred-million dollars? He has five-billion in his slush fund? Look at [U.S. Senator] Barbara Boxer, she sponsored a bill for carbon cap and trade[171]. Who will benefit from hundreds of billions of dollars for administering a scheme like that?

The other thing is research funding. The U.S. spends about two-billion dollars a year on research. Right now, if you submit anything that says CO_2 is not the bad guy, you won't have a chance of getting funding. It all goes to the CO_2 people who build little fiefdoms; they have grant money coming out of their ears. They mimic Al Gore and say the debate is over. The last I heard, the U.S. plans to increase its research spending to 3.5 billion dollars, all of which goes into CO_2 research.

The last part of this equation is the news media and money being made by people like National Geographic who put out a show called Six Degrees of Global Warming[172] and how many people watched that and watched the ads that went with it? How much money did they make doing it? How much money would they have

---

expressed by the IPCC and that all taxes, regulations, and other interventions intended to reduce emissions of CO_2 be abandoned

[171] Sanders/Boxer Global Warming Bill S.309

[172] *Six Degrees: Our Future on a Hotter Planet*, by Mark Lynas

made if they'd said 'Oh, it's not CO_2, it's solar?' Doom and gloom is easy to sell. Herman Goebbels said in World War Two, and said it right, that if you tell a big enough lie often enough, people will believe it eventually.

KLC: With regard to Al Gore's comments, what is your response?

DJE: It reminds me of what went on with the Pope and Galileo. The Pope didn't like the idea that the Earth was round or that it went around the Sun and that the Earth was not the center of the universe. At that point the Pope declared that the debate was over and anybody that disagreed would get burned at the stake. This is like that, total hogwash. He made a statement that less than a half-dozen people in world don't believe that CO_2 causes AGW. He's totally nuts.

KLC: You'd agree there is a small effect of CO_2; the insulating blanket effect.

DJE: That's why we have a nice, warm, cozy planet.

KLC: They're careful in the way they parse words. Everyone believes CO_2 affects our surface temperature.

DJE: That's why we have a warm planet.

KLC: The question is, how much? Is it significant?

DJE: The CO_2 effect is tiny, the eight one-thousandths of one-percent contributed by human activity won't do much. Human-caused warming is dogma, pure and simple. That Al Gore won't debate scientists, won't talk to the press, and all he'll say is 'The debate is over, stupid,' says a lot for the validity of the argument.

There is a list of ten things in his movie, The Inconvenient Truth, that are totally false. I have verified those myself, the assertions are false. To be unkind, they are lies he won't back off from. Hansen, his principal advisor, is upping the ante by saying the sea level change might not be twenty feet, instead, it might be seventy feet. The slides change and they add things on. Al Gore will go down in history as the guy who claimed to invent the Internet and human-caused Global Warming, and they're both bogus claims. It's a hoax, frankly.

KLC: A good friend, Seymour Garte, wrote an optimistic book called *Where We Stand* about the state of our planet. He gives us good news that the air has gotten cleaner, the water has gotten more pure, then he gives you the bad news, I wanted to get your response to this paragraph titled *The Bad News*.

> The major bad news for environmental quality is found in a trend that has been discussed first by scientists and then by activists for many years. Global warming, caused by an excess of carbon dioxide (CO_2) and other gaseous byproducts of industrial human society, has been controversial in the past, but it is no longer. The evidence that we are entering a strong and dramatic period of global climate change has been mounting on a continuous basis to the point where it is now certain. There are still many questions about how bad things will get and how reversible the climate-change trends are. But the fact that the climate is changing is quite real. I prefer the term climate change to global warming because even though it is warming that causes the climate change, we are already feeling the climate change, even if the warming itself is still hard to detect for the average person.[173]

DJE: Certain, ha. The question I have for people like this is, 'please tell me what is the real physical evidence that CO_2 is the cause of AGW?' Tell me. There isn't any. There is no correlation between CO_2 and warming temperatures in the historical record except for the last 30 years which is an absolute coincidence. For 30 warm/cold cycles before that, with far greater amplitude than what we're seeing now, there's no correlation. Why read anything into the last 30 years? It's like saying there's a full moon tonight and it's clear, so the full moon must cause clear skies, right? They occurred together. It doesn't prove anything. There are two lines of evidence

---

[173] *Where We Stand, A Surprising Look at the Real State of Our Planet*, Seymour Garte, PhD.

used to say human-caused global warming is certain. One is that CO_2 has gone up in the last 30 years and so has temperature, but that does not prove anything. The second claim, based on computer modeling, they can reproduce what's happened in the past, and they can project into the future, and clearly show CO_2 as the driving factor. They don't tell you, that, in their computer models, it's assumed that CO_2 drives global warming. In other words, you assume the result and say the computer model proves we were right. It's garbage in, garbage out. If you don't program the computers to cause temperatures to rise with CO_2, then you have nothing.

KLC: It's a perfect circular argument.

DJE: It is. In other words, if it is so goddamned certain, what's the proof? I've asked this question in debates with CO_2 people, and they show the evidence. There isn't any. It's guilt by association. They say the last 30 years proves it. Well, big deal.

KLC: We have fields of expertise. This guy's a good toxicologist, but in his book, he ventured away from his expertise, and instead of exploring the science himself, he takes the things being said at face value.

DJE: I arranged a global warming symposium along with the national meeting of the Geological Society of America in Denver and we invited a half-dozen eminent scientists, all world-class people, to give papers with data to show what's going on. The idea was to get away from the hype and look at the data and see what's going on. When it was all over, one guy stood up and said 'How dare you contradict Al Gore, don't you know the debate is over?' He didn't say son-of-a-bitch, but that was his implication. And another guy stood up and said 'Why are you pointing out all of this data to cast doubt? You're just going to confuse our students.' [Laughs] Guilty.

Anytime you deal with dogma, where people say 'shut up, you can't argue anymore because we're right and there's nothing you can say that will change anything, therefore, you're wrong', attacking the dogma will invariably get you in trouble.

KLC: Here's something new, this is that Six Degrees guy, Martin Lynas. I would like your response. This cute, hand-drawn plot was captured from an ABC news clip, which shows history, today, and talks about the warming pipeline. No matter what we do, we'll get warming to about 2050 where the tipping point appears. The tipping point is the point where actions now can still influence warming at the end of the century.

DJE: The strength of any assertion is only as strong as the foundation it's built upon. This has no foundation. This assumes that CO_2 drives the temperature increase. If it's not, then this means absolutely nothing. You have to prove the basic tenet that CO_2 drives temperature before this means anything at all.

You might as well be talking about the moon being made of green cheese. Until you go there and look at it and see what's there, you can say whatever you want. If you look at the reality of what has happened in the past, which is what I do, and transfer that into the future, you don't get this at all. It turns out that, again, time is on my side, because we're getting closer and closer to my projection and farther and farther away from the IPCC's. We're diverging from this Lynas plot. They predict by 2050 we should be two degrees warmer than today. In three years they say we'll be one degree warmer than today, well, that's not going to happen. This may be an unusually cold year, not necessarily typical of what we have to look forward to, for the simple reason this is a La Nina year, so it probably tacks on a little extra cooling, but the interesting thing is that we haven't had this low of a sunspot cycle since the Maunder Minimum. There are astrophysicists, Russian, Canadian, Willie Soon and other Americans who say 'Look out' because we haven't seen this since 1600 to 1700 and that was accompanied by the LIA, about 4 degrees colder than it is right now. That's one scenario, a possibility.

Another possibility, which I've been speculating on for a while, is the coming cold period that I've projected will be more like the last one from 1945 to 1977, which was half a degree colder than now. Or, it will be more like the one from 1880 to 1910 which was deeper. We might be headed toward one of the deeper ones, like the turn of the last century, but there's no evidence prove it, we'll have to wait until we get there to see. We have the possibility of another LIA. The last time we had such low sunspots; we have the possibility of a cooling spell like 1890, or one like we had between 1945 and 1977, which was mild. Or, we might have nothing at all. Or, we might have soaring temperatures. Of those options, I think we'll have something deep like 1890. But, we don't know and we're not going to know until we get there.

The other thing, which is what the Manhattan and Bali Declarations are all about, is, when we get to 2050, the IPCC predicts two additional degrees of warmth, and the population will increased by up to three billion people. We're projected to be nine billion by 2050. What are we going to do with three billion more people demanding resources? If there is a two degree temperature change, that will be a big problem. So, my view is that, the population is a way bigger problem than a half a degree of temperature change. We need to get control of the population, and we're not.

By the time we get to 2050 we'll have so many more demands for natural resources, even if the two degree temperature change doesn't happen, if we're flat or cooler, we still have a really big problem. Instead of spending two trillion dollars on reducing carbon, which does nothing, we should prepare for the population that's coming. My view has always been that we need to plan ahead. We may have a thirty year grace period when things cool off. If it's only half a degree of cooling, we have breathing space to get ready for what's coming. My projection: we'll be a half a degree warmer from about 2040 to 2070, but the population will be 50 percent bigger, so it's going to be a way bigger problem than what we're looking at today.

Forget about this CO_2 crap. If you bet all of your resources, 2 trillion dollars, on stopping global warming by not putting carbon into the atmosphere, we lose when we get to 2050. In other words, it's a consequential bet. You'll hear the CO_2 people saying 'Well, just to be safe, we got to do it this way', but that's not true because if you put all the money into curbing CO_2 and you don't any of the other things necessary to get ready for increasing population, you have a bigger problem than global warming. That's what the Manhattan Declaration is about. We need to get ready. We should stop pollution. There are far more bad things going into the atmosphere via emissions than CO_2. Sulfur compounds, all kinds of stuff. We'd ought to reduce those. On that, I'm on the same side of the coin as Al Gore.

Suppose that we totally stop CO_2 emissions, take it not to the Kyoto limit but to zero. No heavy breathing. No cars. Would that stop global warming? Will that cause cooling? The answer is 'No.' It takes five-hundred years for the atmosphere to recorrelate with the oceans. Even Hansen will admit that. There's nothing we can do by stopping CO_2 emissions that will affect climate. Nothing. We should bring emissions down, but not for that reason.

KLC: Take Lynas for example. He's says we have to act now with expensive programs, or else.

DJE: The consensus among scientists who think as I do is that the tipping point scenario is invalid. It assumes it is CO_2 causes significant warming. A lot of people will make a lot of money if we do the cap and trade stuff.

KLC: My thought, from talking to you today, is, if we want to do something useful for the climate, we should pray for sunspots.

DJE: That's right, pray for sunspots. You got it. That will work.

KLC: At least the idea won't bankrupt us.

DJE: I see you have a printout of Mann's hockey stick. That is totally discredited now.

KLC: I like the idea someone came up with of 'treenometers'.

DJE: Treenometers? I like that. Al Gore makes the statement that if we go back far enough in time we see the ups and downs in temperature and the CO_2 goes up and down too, which is absolutely true, but what he didn't tell you is, that the temperature change precedes the CO_2 change by 800 years. He makes it sound like the CO_2 causes temperature changes, which is totally bogus.

No geologist, no scientist I know, believes that. But, he's still says it and won't back off on any off it.

KLC: If you want to scare me, show me a plot where we're warm for very little and then cold, during ice ages, for a very long time. That's not good for humankind.

DJE: Actually, it's probably the other way around; there are probably longer interglacial periods. The glacials are probably shorter than the interglacials.

KLC: When can we expect the next ice age?

DJE: There's a really strong correlation between wine production, wheat production, all kinds of things, especially in Europe, correlated to climate which is correlated to sunspot activity.

KLC: Highly correlated to health and societal unrest.

DJE: Absolutely. There is an idea that the French Revolution was brought on by the stress caused by starving people caused by climate change and that's what led to the overthrow of the royal family.

KLC: Let them eat cake...

DJE: Here's a snippet I found interesting. Farmer's Almanac predicted it would be a cold winter? And how do they do that? Do they talk to squirrels or something? It turns out they look at sunspot cycles. Their projections are based on sunspot cycles.

KLC: Online, I said I'd rather believe in Farmer's Almanac predictions instead of the IPCC, at least they have an algorithm, but I didn't know what it was.

DJE: Somebody behind the scenes said 'Yeah, we use the sunspot cycles to make the long term prediction in the Almanac.'

KLC: Looking at this plot, there are regular changes in the ocean level between more than 100 feet higher than now and 300 feet lower than now.

DJE: The rising and falling reoccurs, there is no doubt about that.

KLC: I see areas where the climate is good to us and other areas where it's not.

DJE: These illustrate the Milankovitch cycles, though it was not Milankovitch that originated the idea, it was actually a Scot, James

Croll, Milankovitch applied numbers to the earlier idea, but they're widely known as Milankovitch cycles. Milankovitch cannot explain the Younger Dryas climate changes because the Milankovitch cycles occur on the scale of many tens of thousands of years...

KLC: Yes, about 100-thousand years...

DJE: But in the Younger Dryas period you go from an ice age to a non-ice age in a hundred years and again in a century. This can't be explained by orbital forcing. Milankovitch does not work for the Younger Dryas cooling period so why should it work for any of the others? So you start to think, if it's not Milankovitch, what is it? The first isotope measurements on ice cores in Antarctica showed sudden temperature changes and blew us all away.

Boom, one minute it's hot then it's cold, clearly not explainable by orbital forcing. The idea came up that it all starts in the North Atlantic with the deep current being turned on and shut off. The question is, if this drives the ice ages, quick changes observed in the Northern and Southern Hemispheres should have a lag. It will take 500 to 2,000 years for the effect in the North Atlantic to be felt in the South Pacific or the North Pacific, because it has to go all around the world. You can't transmit across the tropical low of the equator via the atmosphere. If glaciations in the Northern Hemisphere are exactly synchronous with glaciations in the Southern Hemisphere, the North Atlantic current theory doesn't work.

Most of my research has been in New Zealand and South America and various parts of North America and what we've found is that the glaciations are not only synchronous, they're almost exactly synchronous. They are way more sensitively tuned in both hemispheres to occur simultaneously than we ever imagined. It blows the North Atlantic Current theory out of the water which means you can no longer use current theory tacked onto Milankovitch Cycles. It suggests Milankovitch Cycles are not right either. Most glacial geologists would tell you the debate is over, but a number of people doing the kind of research I do that say the facts don't work. What we see is unexplained, so there must be something else.

Yes, there's something else. A growing number of us think the driver is solar. If you plot temperature versus time, it's warm now and 10,000 years back, it was cold. The temperature goes up and down. If we look at the production rate of Carbon-14 and Beryllium-10, it follows temperatures for 10,000 years, not just the last 100 years, or the last 500 years. There has always been an anomaly at the time of the Younger Dryas which lasted about a thousand years and led to a return to the ice age conditions that prevailed for six to seven thousand years. Then it melted like crazy, then it flipped back into cold. The significance of this is when you plot the isotopic values, the isotopes do the same thing and you can't get simultaneous glaciations in both hemispheres if you have to rely on the North Atlantic deep current. The historic climate changes are probably driven by changes in solar activity.

There's no data to make any kind of argument for earlier climate cycles. We need Beryllium. You can get carbon data out of the ice bubbles in the ice cores, but once you get beyond about 100,000 years, there's very little left in the ice core record. When you get back into earlier cycles, you don't have any data. You can't prove anything one way or another. In a nutshell, I don't believe in orbital forcing as a cause of glaciation. Most glacial geologists do, but I don't for the reason I just explained. The data doesn't fit. A growing number realize this.

KLC: I have two stupid questions. I am an electronics engineer, so I know what happens in iron cores, particularly nickel-iron cores, when you have magnetic fields going through them, they self-heat. In fact, it's a very good heater. The Earth is solid in the center, surrounded by a molten layer, and then solid again on the outside. Do we know why the core is molten?

DJE: Put it this way, the basic structure of the Earth with the solid core and the liquid inner core, and then the mantel and crust is based on geophysical data. We think it's true. It's molten because it's hot. Why is it hot? There are two possibilities, one is that it was hot to start with, when the Earth came together and congealed, the hottest part is still in the center, and it's been cooling on the outside. Like a hot biscuit that is cool on the outside, but still hot in the center. The other theory is radioactive heating. It doesn't seem like there is enough radioactivity to do that, so not many people believe that. Most people believe the Earth's core is hot due to a hangover from the process that formed the Earth in the first place.

KLC: Here we are floating in cold space and it's still hot...

DJE: We give off heat as fast as we can, but the damn CO_2 keeps it in. [Laughs] It is a dynamic system, the thought is the thing that causes the Earth's magnetic field is essentially a natural dynamo, there's movement in the core, because its nickel-iron, sets up a magnetic field.

It's a dipole field which, from time-to-time, flips back and forth, probably because whatever is driving the internal convection that causes the magnetic field to form in the first place is not constant. It changes direction. From time to time we have polarity shifts. Most geologists would say the core heating is very likely inherited from whatever created the solar system. Perhaps there were two stars that got close enough so it pulled a lot of stuff out and started it going around the Sun. As you know, all the planets go in the same direction, they spin on their axis in the same direction, which is weird. They all got their momentum at the same time and they're all presumably the same age, so whatever the process was that got all that stuff out there to make the planets, the heating was inherited from that time.

That's the prevailing idea, but I don't know how you prove it. There weren't many people around then. No eye witnesses. No smoking gun. They physics of it are well-known about what's happening now. If you ask if there was ever a time when it wasn't molten, or perhaps molten stuff coming out of the Sun. The difficulty, of course, is the nickel-iron. The Earth has a lot of elements from old stars. The Sun converts Hydrogen into Helium. That's not what the Earth is made out of, so it had to be pulled out of something else, maybe an old star passing by or something. The best answer I can give you to your question is we don't know.

KLC: Where are the places where we're closest to the hot core? Under the oceans. How much do we know about the deep trenches and the interface, could that be a source for heating many times more important than CO_2?

DJE: It's pretty constant, it doesn't change that much. There are certainly hotspots, the Hawaiian Islands are over a hot spot, and Yellowstone is over a hotspot.

KLC: How about some of the deep trenches?

DJE: Those are places where the two plates are being dragged down. There's conduction going on and melting as it gets dragged down. There are places where the plate's temperature exceeds the melting point of rock. All of that is well-documented. When you ask the basic question about why are things not static, why are things the way they are, that's physics. Sometimes the details of the physics are not that clear.

KLC: Dumb question number two is about oil. Where does it come from? I've read some odd theories about it coming from a chemical reaction rather than left-over plant matter.

DJE: Nobody has a clue. We know it's not ground up dinosaurs, which is what you see on the TV all the time.

What do we know about oil? It's always associated with marine deposits, never with continental deposits. It accumulates in areas of porous rock, so it's clear that it migrates. It may not have originated where we now find it. It gets trapped in certain structures. Like any fluid being squeezed through porous media it tends to get stuck in some places where there's a trap or impervious rock that captures it. It can't go anywhere, so it accumulates. If you look at the composition of oil, it's a macromolecule of hydrocarbons.

Where are you going to get hydrocarbons? It sounds like you'll have to have some kind of organic source. You don't normally get inorganic hydrocarbons in large quantities in nature. There isn't any process that will give you that. Most people have thought something happened in the past in the environment that involved a lot of hydrocarbons that got trapped and accumulated and migrated until it's trapped and what we see is hydrocarbons left over from whatever that organic substance was. Beyond that, we don't know. If you want to be rich and famous, find out where does oil come from.

KLC: Here's a chart that shows ocean levels spend a lot of time 300 feet lower than they are now and 100 feet higher than they are now. Does this match what you know about ocean levels?

DJE: Oceans rise and fall depending on how much ice there is on land. There's a clear connection. In order to get significant change in ocean level, you have to tie up a lot of water in ice on the continents. The continental glaciers covered almost all of North America, clear down to Ohio, Kentucky and it was like the Antarctic ice cap, it was 10 to 15 thousand feet thick. Huge volumes of ice. The same thing in Europe, the same thing in Russia, Siberia. If you melt all of that ice, you get about a 120 meters of sea level rise. That's the difference. The fluctuation of 100 meters of sea level can be explained very nicely by ice volumes found during the ice age. People made calculations to see how much water was tied up in the glaciers, and what that should do to sea levels and so on. In terms of amount, a sea level change between 20 meters and 100 meters is what you'll get during an interglacial period.

KLC: There's so much talk about sea level changes, particularly increasing. Where are we in the one-thousand or five-thousand year cycle?

DJE: Nobody knows. There are suggestions. If the main driver is solar, there are long term solar cycles, but the trouble is long term solar cycles are hard to find because they don't leave evidence behind, unless it's climate.

Then you argue: 'How do you know the root cause is solar?'. The lesson here is, until you solve the problem of what causes the ice ages, you won't be sure about where we're headed in the future. In my opinion, the odds are that the climate changes are driven by solar influences because of the isotope records, since correlation doesn't work for Atlantic currents and Milankovitch Cycles. If recent climate changes aren't caused by continental drift or deep ocean current or solar radiance, then what is it? You run out of options. It used to be that physicists would tell us that solar output was immutable, it never changes, forever. Now physicists say they were wrong and solar constants aren't, they really do change. You can get good correlation to 11-year sunspot cycles and radiance. They are right on top of each other. They follow each other beautifully. When is the next ice age coming? Who knows?

KLC: For the record, Don, if you gave a gift to the world and controlled the thermostat, where would you place us with regard to climate history?

DJE: If I had to pick a climate that was good for the whole world, the perfect climate would be around the Medieval Warm Period...

KLC: Which is similar to where we are now...

DJE: We're slightly below it. It depends on who you believe. When the Mann hockey stick curve was all the rage, they said the MWP didn't happen, now we know it did. My work shows it. During this time, civilizations flourished in Europe because of the long growing season and other things. There is good reason to believe if you're in that range, the growing season is longer, you can grow more food, you can grow more food in Northern latitudes, and you'll support a more robust civilization. People will have more free time because they're not starving to death, they can do more things like art...

KLC: And study the climate...

DJE: Right, like study the climate, that sort of thing. If I had to pick a climate that would be a nice thing for the whole world, I'd say somewhere close to the MWP.

KLC: A little bit warmer than now...

DJE: Yes, a little bit warmer than now, but not much. There's an interesting parallel. If you look at the temperature curves, we've been coming out of the LIA for about 400 or 500 years at a rate of a degree a century. Will we do this forever? A degree a century? We have the thirty-year wiggles in there, but when do we top out and start cooling again? During the early part of the Holocene, it was warmer than now. In fact it's called the Climactic Optimum; it was warmer than it is now. If we are on an overall rising temperature curve coming out of the LIA, when we get to the temperature of the MWP, will we get another LIA, something really big or something in-between like the 800-year cooling? The answer is, we don't know, we'll have to wait and see. Until we have a better understanding of what causes climactic fluctuations, solar or not solar, and if it is solar, what the impact of solar fluctuations are, there are a lot of things we don't know. Until we understand the mechanisms better, we don't know. I don't know. I don't know anybody who does know.

KLC: I had a bias going in. I didn't believe any of the AGW-crap, but I was challenged to study and follow the arguments. I said, 'I'll open my mind as much as I can, see what they're saying, and accept that maybe I'm wrong in my bias.' But, that's not the conclusion I came to.

DJE: If I were hard-pressed to give my overall assessment of the whole Gore-phenomenon, I would say two things, A, that it's an out-and-out hoax, and B, it is probably the biggest scientific boondoggle since the days of Galileo. There is so much dogma and pressure put on scientists. Gore has so little proof, it's really sort of disgraceful to the scientific world.

KLC: I think it's more than that. There is self-loathing, they believe that man's works are inevitably bad, so they contort to prove that result.

DJE: Follow the money. It's big bucks. We're talking about billions of dollars.

If we're headed toward catastrophic global warming, we have to do all these things they want. It will prove to be the biggest boondoggle since Galileo.

Ken Coffman

ARGUING WITH MORONS

All you need in this life is ignorance and confidence, and then success is sure.
—Mark Twain

THERE ARE MANY famous quotes about the fruitlessness of arguing with morons, but I find it entertaining on occasion. I hope the patient reader won't think I claim to be always right, because I'm not—I'm a buffoon and I've been proven wrong many times.

Provided here as a cautionary example of arguing with dimwits, let's look at a conversation I had with my brother, Ike, on Facebook.

Ken Coffman: Let's see the paper that proves HIV causes AIDS or if you want something easier, how about the evidence supporting the idea that the GHE increases the Earth's surface temperature by an average of 33°C. This number is in many physics texts—what are the mechanics? What is the distribution? ±10%? ±50%? ±100%? Is the distribution Gaussian? Is it the same at the poles and at the equator? Is it the same in winter and summer? Good luck.

ILC: Ken, I am way too busy to do your research for you. Really, all you have to do is type in phrases like "does global warming exist", or "does HIV cause AIDS", and you will get tons of evidence. However, it does truly surprise me that you would rather believe paid industry shills from places like the Heartland.org or wackos like Bauer than from sources like, say, the EPA, NASA, NOAA, (for global warming), or from the CDC or WHO for HIV. And where in the world do you get that 33 degree C number? The Earth has warmed slightly less than 1 degree C, with the "threat" of an increase of 5-8 degrees, or so.

Okay, I can do some research for you. Start here: The Conversion of a Climate-Change Skeptic.[174]

KLC: Ah, the appeal to authority argument. Muller is an academic shill and I say that though his critical thoughts about Michael Mann's iconic hockey stick graph were accurate. As always, your thoughts will be influenced by who you trust and I don't trust Muller. I'd take Dr. Easterbrook's word over Dr. Muller's all day long. I don't expect you to change your mind. I don't want you to change your mind. Go ahead—wallow in ignorance and enjoy its fruits. Don't even try to figure out why you believe certain things—that would take a level of insight and objectivity for which you are incapable.

The 33°C of GHE warming is repeated over and over and it doesn't take much thought to realize how stupid the idea is. Regardless, here is some of the typical drivel:

> Greenhouse gases let the Sun's light shine onto the Earth's surface, but they trap the heat that reflects back up into the atmosphere. In this way, they act like the insulating glass walls of a greenhouse. The greenhouse effect keeps Earth's climate comfortable. Without it, surface temperatures would be cooler by about 33 degrees Celsius (60 degrees Fahrenheit), and many life forms would freeze.[175]

ILC: Are you saying you believe that the greenhouse effect is drivel?

---

[174] http://www.nytimes.com/2012/07/30/opinion/the-conversion-of-a-climate-change-skeptic.html?pagewanted=all&_r=0

[175] http://education.nationalgeographic.com/education/encyclopedia/global-warming/?ar_a=1

Wow. You do understand that some light frequencies that can pass through the atmosphere get absorbed by the Earth, then re-radiated at a longer wavelength that IS absorbed by the atmosphere? Right? If that effect is 33°C, then how is it not reasonable that an increase of about a third of the original amount of greenhouse gasses can raise the temperature another 8°C? Or are you going to deny other scientific theories, like gravity?

KLC: This argument strategy is called "straw man" where you build an artificial, invented point and then demolish it. Even some rabid global warming activists have given up on the thought that greenhouses, which work by restricting convection, work by restricting radiation.

I won't argue about the semantics—you can erroneously call it the GHE (Green House Effect) if you like. There is a lovely asymmetry to how we take in high intensity radiation and emit low intensity radiation which is a warming effect, but has nothing to do with the atmosphere and particularly nothing to do with trace constituents of our atmosphere.

I know there are the equivalent of Fraunhofer lines in our outgoing longwave radiation caused by water vapor, CO_2, methane etc., but the question is, what work does this do? Does it increase the Earth's average surface temperature? The progressive activists have no idea how useful it would be to increase the average temperature of something, like your living room. Or, to intensify this effect and increase the average temperature by 66°C. Or to block this effect and cool something by 33°C. This would be hugely useful—and to build a machine like this with cold, thin air? Wow! Gimme some of that.

ILC: You start off by calling my statements a "straw man", then you create one yourself. It is obvious to me that you fundamentally misunderstand the science. You know some terms (Fraunhofer) but you apply them incorrectly. Yes, a real greenhouse and the Earth's atmosphere greenhouse effect work differently. We use the term because they both create a warming effect. Look, I cannot give you a complete science class here on Facebook. Just go to Wikipedia and look it up: Greenhouse Effect.[176]

KLC: Ah, now we're back to an appeal to authority and the authority is Wikipedia, a horribly corrupted and hopeless compendium of nonsense when it comes to global warming. I have invested the 10,000 hours required to become an expert in electrical engineering and associated thermodynamics, but you can put your full faith and confidence in William Connolly and not me. Have at it. Take up a collection and build a statue of Connolly in the park, I don't care.

ILC: I am not appealing to authority. I am saying I understand the science and the process, and that you do not. I am not going to try to explain it to you, you can either do the research yourself or not.

However, it makes you look absolutely foolish to deny something that you obviously do not even understand. Listen, this is basic science, and it's easy. The science behind global warming is what also allows current to flow by the creation of free electrons through energy absorption. If you are saying the science behind global "climate change" is wrong then you are saying the science behind electricity and heat transfer are also wrong. You are also saying that the entire rest of the world is wrong, and that there is a vast conspiracy to hide the truth for some nefarious reason. It simply boggles my mind.

Can you not believe that perhaps, just perhaps, scientists are concerned that we are destroying ourselves, and that they feel like it is part of their job to at least try to warn us?

KLC: Instead of broadcasting ignorance (and another dumb straw man argument, by the way), why don't you do something useful and study Figure 1 of the Trenberth/Fasullo/Kiehl link below[177]—a schematic almost as iconic as Mann's corrupt hockey stick graph. Engage the engineering part of your brain—and disengage the blind

---

[176] http://en.wikipedia.org/wiki/Greenhouse_effect
[177]

http://www.cgd.ucar.edu/ccr/aboutus/staff/kiehl/EarthsGlobalEnergy Budget.pdf trust in authority endemic in second-handers like you.

ILC: It appears to be legitimate—and important—research into the mechanisms and relative importance of the factors that maintain our energy balance. Virtually all systems maintain a balance because there are self-correcting mechanisms that occur naturally.

Take a system you know very well: a power supply heat sink. Most of the time, heat in equals heat out. The important mechanism is that if heat in increases heat out will also increase, but it requires a higher temperature to do so. You know, the old heat transfer equation, q equals m c delta T.

However, nothing in this paper is an argument either for or against global warming. And in fact, the authors state that many of the variables cannot be measured accurately. This paper is important because it gives us a baseline to see if changes are occurring in those variables, and it gives us a starting point to try to measure those variables more accurately. The important question— one this paper does not attempt to address—is what is the sensitivity of Earth's temperature to an increase in heat retention. We know the Earth is pretty darn efficient, but the fact remains that we are retaining heat.

KLC: Clearly, even the smallest amount of intellectual curiosity is asking too much. "The fact remains that we are retaining heat." Hmmm, interesting, retaining heat. Retaining heat in what? Cold, thin air? Think about it.

ILC: Ken, you yourself mentioned the term Fraunhofer. Yes, cold thin air. Air that has more carbon dioxide will absorb more radiated heat (longer wave light, essentially) than air with less carbon dioxide.

If even one photon that would have ordinarily zipped off into space (at the speed of light, I might add) gets held by the atmosphere for even a nanosecond instead, the Earth has now increased its net energy by that amount. But you are right, that amount may really be inconsequential. The real problematic factor is when that photon (which would have zipped off into space) gets released by the carbon dioxide molecule (which it will do, and relatively quickly) back in the direction of the surface of the Earth, where it gets re-absorbed. Net energy gain, now a substantial one.

KLC: I only care about transactions near sea level where the atmosphere is densest—in this area, a dissipative collision is around 30X more likely than a photonic re-emission. For these rare emissions, I have no problem with the idea that the outgoing photon is translated into an omnidirectional emission, but you have to ask yourself, given that a heated (or overheated) CO_2 molecule will convect outward with mad abandon, what is the overall effect? I tip my hat to you for recognizing the effect of CO_2 must be small, but you must also recognize that the overall small effect of adding a CO_2 molecule to our atmosphere might be cooling.

The progressive activists love playing tricks with radiation like it's a trained circus animal jumping through fiery hoops—and they love talking about mysterious radiation modulators implemented in cold, rarefied air—modulators capable of amazing magic tricks, but anything the upper atmosphere can do is heavily filtered by what happens near the Earth's surface. In no possible scenario are GHGs capable of increasing the Earth's average surface temperature by 10% (33°C, or better, 33K).

Think about all the different ways air could heat water. Of all these collected methods, which are in effect in our cold, thin atmosphere?

ILC: Dissipative collision? I don't understand that. Collisions will transfer energy, but energy is never dissipated.

KLC: Really? You understand temperature is a facet of motion or vibration. Once something is in motion or vibrates, it always stays in motion or vibrates? Interesting.

ILC: I thought we were making progress, now we are going backward. Yes, unless the atom/molecule is at absolute zero it will always be moving/vibrating AND colliding with other atoms/molecules, which transfers energy but does not (cannot) dissipate it. Energy is neither created nor destroyed, but it can change form. The lack of movement/vibration is kind of the definition of absolute zero, which is really more of a concept than a reality because we can never reach absolute zero for the same reason that mass will never travel at the speed of light. Back to the point, dissipative collisions DON'T dissipate, they transfer, and we are still dealing with the reality that an increased carbon dioxide level increases the amount of energy the Earth must absorb.

KLC: Here are synonyms for dissipation: disperse, dissolve and scatter. Near-visible, infrared, photon-carried energy does all these things as it is emitted from water and Earth. It dissipates into the air. It's integrated. Dispersed. Spread out. Diffused. Refracted. Reflected. Absorbed. You should probably spend your time talking about sports or something safe.

ILC: None of that makes the energy go away. Every single one of those terms is a transfer of energy, which is what I said. Just because the heat spreads out doesn't mean it isn't getting hotter.

KLC: If you're ever interviewing for an engineering position, here's a helpful suggestion: don't say "Just because the heat spreads out doesn't mean it isn't getting hotter" to the hiring manager.

ILC: We are talking about global warming, not a power supply. In a power supply the heat that is "dissipated" really gets spread out into the rest of the room, which is fine. Power supply temperature goes down 100 (or so) degrees, room temperature goes up .0001 degree.

No big deal, this is how a power supply works. It is a big deal in a server farm, where you have so much heat it raises room temperature so much you need special cooling. For the entire Earth, there is no "special cooling", you seem to think that somehow (must be magic, or something) dissipation is that special cooling. It's not. The Earth has warmed up .7 to .9 degrees C because the heat retained by carbon dioxide doesn't (and can't) get dissipated.

Instead of arguing about dissipation, just tell me where you think the heat that gets dissipated goes. I get the feeling you think I am just trying to shoot down all of your theories. I am not. Reality is.

KLC: All passive heat is headed in the same direction, toward absolute zero. The Earth is heated by the Sun and eventually all that heat energy will get to where it would be if the Earth did not exist—dissipated in empty space.

ILC: Simply not true. Here is where we get to the crux of the matter: you can either learn how the universe actually works by studying the science, or you can continue to believe things that are not true. Or, if learning the science is too much work, you can accept what the people who DO know the science are telling you.

KLC: You should take up climate science where a pathological disconnect from physical reality is no impediment to success. Like the academic deep thinker Kevin Trenberth in the Figure I referenced above, if you want to believe the Earth's surface is heated more by atmospheric "back-radiation" than insolation, then have at it. I like competing with guys like you in the job market.

ILC: Not even space is at absolute zero. There are only 3 ways heat can move, and those are conduction, convection, and radiation. Conduction works when molecules or atoms collide, which requires matter, and that method works on Earth but not in space. Convection is a flow of matter (usually liquid or gas), and also does not work in space.

The only way energy (heat) is transferred into (or through) space is through radiation, which is how we receive energy from the Sun and also the way the Earth gives off energy to maintain an energy balance. When some of that radiation is blocked from leaving the Earth by stuff that absorbs it (carbon dioxide) the total amount of radiation has to increase in order to maintain that balance.

Radiation increases when temperature increases.

It REQUIRES a higher temperature to maintain an energy balance when the Earth retains more energy. The increase in temperature will naturally occur in any system that has more energy coming in than going out. When energy in increases, there is a temporary imbalance, temperature goes up, the system starts giving off more heat, and the system gains a new energy balance AT A HIGHER TEMPERATURE.

Climate science is not what has a pathological disconnect from reality. Geez, Louise, give a conservative some facts and he just starts insulting you. Typical.

KLC: Radiation is not blocked by atmospheric gases—it's not trapped, stored, retained or impeded. It will get to nearly-empty space. The only thing the atmosphere can do is delay the pace of exiting energy.

The question is how long the delay is. It will depend on the path, so there is a distribution—the energy that exits in atmospheric "windows" leaves our system at a significant percentage of the speed of light, but wavelengths that interact with resonant molecules are delayed by long periods of variable time. However, once you realize the median delay is on the order of 3-7 milliseconds, then you begin to see the light.

There are ways to store thermal energy for long periods of time, but these methods have nothing to do with rarefied CO_2 and cannot be implemented in a cold, thin atmosphere. In order to create positive feedback in the diurnal (day/night) cycle, the storage and delay periods have to be many hours in length. This is why climate activists immediately reach for integrating tools and talk about averages—average areas and average energies. It's hand-waving bullshit and you'll see it every time a climate activist tries to explain in detail how human-caused global warming works.

I spent a decade studying this topic and I understand Warmista arguments better than they do. I understand the analysis, but I don't agree with science. Post modern science really offends me. You don't have to believe me. In fact, I prefer you right where you are. You don't believe a huge herd of climatologists could be so off base. So deluded. So wrong. But, I'm telling you, NGO and government-funded activists are hopelessly corrupt and yes, pathologically disconnected from physical reality. As an engineer, if I want my electronics, currently flying around on airplanes, to work reliably, I have to work in accordance with the laws of physics. Because I'm not an academic, I can't make up things and expect the universe to conform to my desires.

ILC: Trapped? Stored? No. Retained? Impeded? Yes. The radiation will eventually get to space, and you are correct that the only question is how long the delay is. It is not the energy delayed by 3-7 ms, it is the energy directed back at the surface of the Earth (that would otherwise not have been). Tell me exactly what the delay time of that energy is.

KLC: Most of the radiated energy leaving the Earth's surface escapes quickly at some large percentage of the speed of light. There are narrow bands of wavelengths where complex molecules resonate and turn unidirectional photons into omnidirectional photons, but molecular energy is either dissipated via collision or quickly re-radiated as another omnidirectional photon.

The path length for photons is variable—it depends on how much they bounce around—and you could make a case that an occasional photon will bounce around forever, but the energy value of those photons is insignificant and irrelevant. The median delay period between photonic energy being emitted by the surface and being lost to space is about 5 milliseconds.

ILC: Nothing about this is postmodern. This is all about well known, time tested science. The problem is that the Earth and the atmosphere are complicated, dynamic systems. We still have not figured out all the dynamics to this day, and that is certainly part of the reason it seems more complicated than it is.

We understand the basics of the mechanism, we do not understand the relative amounts.

It is a lot harder to measure historical temperatures accurately, something you know (the hockey stick), but it is a lot easier to measure historical carbon dioxide levels. We know the level of carbon dioxide has NEVER been above 300PPM, going back as far into history as we can (thousands of years). It is now at 380PPM, and climbing. CO_2 does absorb infrared light energy, more commonly known as heat. More CO_2 absorbs more light. Much of this light is re-radiated (after the 3-7 ms delay) into space, but other amounts get radiated into direction where it is absorbed by other CO_2 molecules, and a smaller fraction gets radiated back toward the

Earth. It is these relative amounts that have not yet been accurately calculated, and where the Trenberth et al study is so important to set a baseline. But that it happens, that it is increasing, and that it is raising global temperatures in not in question. Those facts have been established.

KLC: "We know the level of carbon dioxide has NEVER been above 300PPM, going back as far into history as we can (thousands of years)."

Would you like to rephrase that? (See Figure 11)

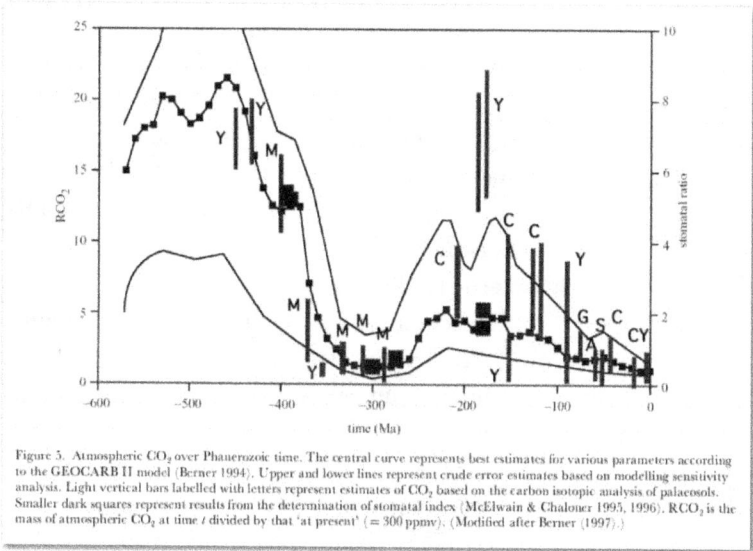

Figure 5. Atmospheric CO₂ over Phanerozoic time. The central curve represents best estimates for various parameters according to the GEOCARB II model (Berner 1994). Upper and lower lines represent crude error estimates based on modelling sensitivity analysis. Light vertical bars labelled with letters represent estimates of CO₂ based on the carbon isotopic analysis of palaeosols. Smaller dark squares represent results from the determination of stomatal index (McElwain & Chaloner 1995, 1996). RCO₂ is the mass of atmospheric CO₂ at time t divided by that 'at present' ($= 300$ ppmv). (Modified after Berner (1997).)

Figure 11

Historic Levels of CO$_2$[178]

[...]
Denis Rancourt's conclusion[179] about the human-caused global

[178] http://shadow.eas.gatech.edu/~jean/paleo/Lectures/Lecture_3.pdf
[179] In 2005 and 2006, several years before the November 2009

warming theory is 100% correct, so he'll be my friend forever.

ILC: A blanket statement like that calls into question not only your intelligence, but your sanity as well.

KLC: I made several statements—some are opinion and some are factual. Let's zero in on the specific blanket statement you disagree with.

ILC: The blanket statement I was specifically referring to is the 100% comment. I am not 100% sure that global warming is real and is caused by humans, but the evidence is both real and overwhelming.

Even more damaging, looking at the anti community in general, it is filled with wackos who lack credibility. Their science is not sound, they cherry pick data, they engage in behaviors and analysis that is exponentially more misleading that what they accuse the pro side of engaging in. To believe in the anti global warming rhetoric is to reject sound science and embrace some sort of global conspiracy theory. It's just not rational.

KLC: I know it's hard to believe the established authorities demonstrate such a bizarre combination of corruption and stupidity, but here we are. Once you grasp how wrong-headed the "consensus" global warming argument is, it makes you question all the conventional wisdom.

I know you, thoroughly indoctrinated and conditioned, accept what you're told with religious conviction, but just engage your brain for just a few minutes to consider how 400PPM of a benign, life-essential gas can warm the water that covers 71% of our planet.

Imagine an IR-resonant CO_2 molecule dancing in front of your

---

Climategate scandal burst the media bubble that buoyed public opinion towards acceptance of carbon credits, cap and trade, and the associated trillion dollar finance bonanza that may still come to pass, I exposed the global warming cooptation scam in an essay that Alexander Cockburn writing in *The Nation* called "one of the best essays on greenhouse myth-making from a left perspective."

—Denis Rancourt, *Hierarchy and Free Expression in the Fight Against Racism*, P. 115 Stairway Press, 2012 face. When it gets stimulated, it convects upward and becomes an agent of cooling, not warming. Sure, it radiates omnidirectionally and thus is also a weak agent of warming, but what is the relative proportion? What are the necessary conditions for radiation to trump convection? Are these conditions present anywhere on the Earth's surface? Good luck. Have fun.

It's interesting to me—there is a type of person who will believe theory over the evidence before their very eyes. For example, on my walk yesterday, I noticed ice on the boardwalk—melted where the Sun directly heated the wood and still frozen in the shadows. Interesting how that could happen when, according to Trenberth/Kiehl/Fusillo, the "back radiation" power from the atmosphere is roughly 2X the average power from insolation. Sure, the insolation is "peaky" and heats during less than half the 24-hour period, but how can the integrated effect of the back radiating atmosphere contribute 2X the heating power of the Sun? That in a nutshell is post-modern science, my friends.

RANDOM INTERVIEWS

All methods strongly underestimate the amplitude of low-frequency variability and trends. This means that it is almost impossible to conclude from reconstruction studies that the present period is warmer than any period in the reconstructed period.
—Bo Christiansen, Danish Meteorological Institute

Claes Johnson

HERE IS A short interview I did with mathematician Claes Johnson—who is simply a brilliant man.

KLC: First of all, from a purely radiative point of view (with no conduction or convection), is it possible for a passive radiator to increase the peak or average temperature of an active radiator? (An active radiator is one with an internal source of heat energy).

Claes Johnson: As posed, I cannot answer the question, because my answer must be: 'It depends…'.

As a start, let me clarify my standpoint. Heat transfer (or heat flow) by conduction and radiation is governed by temperature gradients described by Fourier or Stefan-Boltzmann laws, with the flow of heat in the opposite direction to the temperature gradient,

Ken Coffman

which effectively seeks to decrease the gradient by taking heat from the hot (rich) and giving to the cold (poor).

Heat transfer by conduction and radiation is thus a diffusion process tending to decrease temperature gradients (differences) by smoothing or averaging. This process of (positive) diffusion of taking from the rich and giving to the poor—decreasing wealth differences—can be compared with the reverse process of negative diffusion of taking from the poor and giving it to the rich which increases wealth differences.

Now, a process of (positive) diffusion process is stable (Sweden) and is realized in the physics of heat transfer by conduction and radiation. However, a process of negative diffusion is unstable and breaks down (Egypt) and does not persist over time.

The above concerns heat transfer without forcing. With forcing anything can happen. By forcing (by a compressor, for example), heat can be transferred from the cold inside of a refrigerator to the warm outside. But if the compressor is shut off the warm exterior will heat the cold interior, inevitably and always.

With regard to radiation, it remains to prove Stefan-Boltzmann's Law of heat transfer from a warm blackbody to a cold. The standard proof is by statistics of quanta, which is difficult to understand and is questionable from physics point of view since there does not appear to be a department of statistics inside blackbodies nor between them.

I give a different proof based on a deterministic wave model subject to some form of finite precision computation, in analog form in real physics, and in digital form (finite number of decimals) in computational simulation. I show that a warm body with a spectrum containing high frequencies will heat a cold body, because a cold body cannot emit high-frequencies (because its "lips are too stiff"), but cannot protect itself from absorbing coordinated high frequencies in resonance, and the only way out is to break down and store these incoming high frequencies as uncoordinated high-frequencies (i.e. heat). In this model heating comes from "eating" high frequencies which cannot be directly re-emitted, and thus a cold body cannot heat a warm body because the "food of high frequencies" is missing.

Again, what is said above concerns the interaction between blackbodies without forcing. With forcing—as in a microwave oven—anything can happen, i.e. forced heating can occur via high-amplitude low-frequency waves.

Ken Coffman

Radio host Dennis Miller interviews Sky Dragon Slayer Joe Olson...

Joe is a co-author of *Slaying the Sky Dragon—Death of the Greenhouse Gas Theory*.[180]

Dennis Miller: We're going to go now to our guest, he's Joe Olson, he's one of the authors of the Book *Slaying the Sky Dragon—Death of the Greenhouse Gas Theory*—a series of essays, I guess. Joey, welcome to the show.

Joseph Olson: Hi. You're aforementioned Miller-man?

DM: That's what they say. I was before—global warming affected me.

JO: Host of the thinking man's radio...

DM: Listen, I don't know about that. Now don't start jerking me around here. But, I have never believed in global warming. Did you at one point, and you fell off? Have you never bought it? I'm documented on HBO specials for years now making fun of the concept. What's your take?

JO: Well, actually, I'm the science guy without the bowtie and this thing never made any sense to me from the get-go. But, until they started to be really threatening from a governmental level here in America, I pretty much ignored it. But I'm opposed to the faux-science witch doctors of climate change.

DM: The smartest man I ever met did not subscribe to it. That would be, of course, Michael Crichton, and when we talked he did not subscribe to it and that piqued my interest. I love *State of Fear*[181] and thought the annotation at the back of the book was very interesting. Tell us about your book *Slaying the Sky Dragon—Death of the Greenhouse Gas Theory*, how was it composed?

JO: It's available at Amazon, Barnes and Noble and StairwayPress-

---

[180] *Slaying the Sky Dragon—Death of the Greenhouse Gas Theory*, Stairway Press, 2011

[181] Michael Crichton's 2003 novel debunking human-caused global warming.

dot-com. It's a compilation of the works of eight different authors. I'd like to run through a brief little outline, quick, I've got a hundred articles posted at Canada Free Press[182], The Freeman Institute and Climate Realist.[183] I've been cross-linked to two-hundred-thousand websites in a dozen languages and read into the congressional record. There're three things about global warming that are glaring errors, one of them is thermal mass, one of them's infrared transmission and one of them is CO_2 toxicity. So, we can get into those if you want to or we can discuss the composure of the authors.

DM: Let's talk about the three things, you're right, because the author's names probably won't mean much—but do it as 'laymanesque' as you can. While I don't believe, also I'm not a wizard.

JO: Okay. Well, first of all, thermal mass. You can take a red-hot BB and drop it in a swimming pool and they're both going to reach the same temperature rather quickly as a function of specific heat, mass and difference in temperature. But I guarantee you, the BB will not warm the pool up much...

DM: I learned that one night at Axel Rose's house.

JO: [laughs] So anyway, what humans have done is put twenty-eight giga-tons—that's tons with nine-zeroes behind it—of carbon dioxide in the air. Carbon dioxide is a three-atom molecule—it's like nano-dust. It's plant food. Below three-hundred parts-per-million, plants atrophy, below two-hundred-and-fifty parts-per-million, they die. We currently have three-hundred-and-ninety parts-per-million—so that's like a minimal level.

DM: Yeah. Negligible.

JO: Yeah. And the carbon that's in the air will be absorbed by plants and something called the *carbon cycle*—which *carbon* life forms depend on. And it will turn into dirt or it will be absorbed in the ocean and it will be absorbed by marine animals and turned into

---

[182] www.canadafreepress.com
[183] www.climaterealist.com

calcium carbonate which we like to call $CaCO_3$. Dirt and calcium carbonate have a weight of a hundred-and-twenty-five pounds per cubic foot...

DM: See, you're losing me now, Joey. I'm just jumping in and telling you, if you're losing me, then I think you're losing some listeners...

JO: Just quickly...

DM: But I'm telling you, you can take your time and lose me here, but I'm starting to get lost. Go ahead.

JO: The twenty-eight giga-tons means less than three cubic miles on a planet with two-hundred-and-fifty-nine trillion cubic miles of molten rock—or three-hundred-and-ten million cubic miles of ocean. So basically, the carbon dioxide is the BB and the planet is the swimming pool.

DM: Now I get it again. That's what I need. If you said that to [Al] Gore and he was in one of his wild-eyed manic things, how does he counter that? What would *he* say? What would Gore say? Gore representing the other side. What would they say? That you're just lying? Your facts are wrong? I'm always intrigued—when presented with empirical evidence, that's not enough for them, right? They're on the ledge and they ain't comin' in.

JO: Well, yeah, they're invested in creating another commodity market and so, they don't have to have any scientific basis.

DM: I do believe that. They're into the 'we-have-no-future' futures—is I think what they're selling.

JO: Yeah. It's the credit-default swap of science is what they're doing.

DM: [laughs] The C-D-Os [Collateralized Debt Obligation instruments]. I just read *The Big Short*,[184] so I'm actually hip to that. Tell me the second thing—the second tine on your fork.

JO: Yeah, the infrared transmission. The Earth absorbs a full spectrum of sunlight during the day, then at night it re-radiates it. Carbon dioxide, because it can only absorb in certain spectral

---

[184] Michael Lewis, *The Big Short: Inside the Doomsday Machine*

ranges, absorbs in the five and fifteen micron range. One of those it shares with water vapor, so if there is any water vapor in the air, the amount of carbon dioxide doesn't matter. There's only a limited amount of infrared radiation being given off from the planet, so the more carbon dioxide you add, it can't add additional infrared radiation, so it can't capture any more than a finite amount anyway.

DM: So, on a planet that's [covered with] eighty-percent water with a Sun that heats it, there's always going to be water pressure—or water vapor in the air.

JO: You hope. If there's not, it's a desert.

DM: [laughs] There you go. If not, we're Charlton Heston in the future—we've just plunked it into a sandstone or something.

JO: But what they don't tell you is the infrared radiation is electromagnetic, it travels at the speed of light, so this energy is leaving the Earth at a hundred-and-eighty-six-thousand miles per second. It bumps into some carbon dioxide atoms [molecules], but the lapse time that it stays in each carbon atom [molecule] is less than a billionth of a second. So, by the time you bump into a few of them on the way out? You've slowed that energy down by twenty milliseconds.

DM: Well, listen, I can tell you, to quantify that figure for the folks out there in the listening audience, if they pass cap-and-trade—that's only a thousand miles-per-hour less per second than I'm going to leave the freakin' planet. I'm not paying it. Third thing.

JO: The third thing is CO_2 toxicity. NASA, when they were getting ready to put a fire-suppression system in the space shuttle and in the space stations—tested carbon dioxide—they saw no measurable side-effects in concentrations less than eighty-thousand parts-per-million. The air has three-hundred-and-ninety parts-per-million. We inhale that three-hundred-and-ninety—we exhale forty-thousand parts-per-million.

DM: But can I ask you the layman's question? How many parts per million were they about to get in the movie[185] when they were

---

[185] Apollo12

going to die from CO_2 poisoning? Didn't they have to do an air cleaner or something?

JO: It's absurd. You're in buildings all the time with two-thousand parts-per-million.

DM: Okay. I didn't know—I'm just asking. You're the smart guy...I'm the dumb comedian.

JO: Every windbag on the Supreme Court creates a hundred-times more CO_2 than they inhale.

DM: [laughs] I like it when you get to the latter part of your things and tie them together. They make more sense, 'cause I glaze over during the number part.

JO: Back to Slaying the Sky Dragon...

DM: And the website, by the way, is SlayingTheSkyDragon-dot-com.

JO: That's one of them. Yes. The collective effort of eight scientists from five countries who met on the Internet—one of them is a PhD Climatologist,[186] one's a PhD Physical Chemist,[187] one is a PhD applied mathematician,[188] one's a PhD Physicist.[189]

DM: Don't give me all eight here—eight smart guys. They meet on the Internet? Where? Craigslist or something? Where?

JO: No, we started writing articles—I have like a hundred articles posted at various websites—sixty at Canada Free Press, and in the process of reading each other's work, we'd send each other a message. "Hey, great article."

DM: Yeah.

JO: One of us said, "Hey, why don't we get together and write a book?" The next thing you know we were writing a book and none of us had ever met. We're from five different countries.

DM: Well, you got a wizard club there—now, listen, let me say this—and the book, once again, *Slaying the Sky Dragon—Death of the Greenhouse Gas Theory*. We're talking to Joe Olson. Whoever did the

---

[186] Dr. Tim Ball
[187] Dr. Martin Hertzberg
[188] Dr. Claes Johnson
[189] Dr. Charles Anderson

cover art?[190] It's great. It should be on the side of a van, it's so cool. Who did the cover art on the book?

JO: It was done by two different people—the English version was done by John O'Sullivan—who's one of the authors. The American version was done by Ken Coffman with Stairway Press. Yeah, it's great cover art.

DM: It's great looking. Thank you, Joey, for your time.

JO: Hey, appreciate it.

DM: John O'Sullivan must have been feeling his oats after beating up Gentleman Jim Corbett here—because this is a nice drawing and we appreciate Joe Olson's time. And, I'm getting Sun flares that are breaking up my connection, everybody. No, I'm not, I'm just kidding. Everybody try to cool out and enjoy your Thursday—forget the planet for a day. Back after this on the Dennis Miller Show.

---

[190] Guy D. Corp, www.grafixcorp.com

Ken Coffman

BIBLIOGRAPHY

IF YOU LOOK carefully at the back cover, you'll see a picture of a partial stack of books cluttering up my kitchen; these are books from my personal library that I read and reread—looking for evidence to support the idea that adding CO_2 to the Earth's atmosphere increases its average temperature. If I missed a key corroboration in any of these books, please let me know the chapter and verse. Thank you.

I am not smart enough to discover the inner secrets of thermodynamics on my own—everything I know came from books, online studies and conversations with experts. They deserve all the credit for any accidental wisdom I achieved.

These books are listed in random order—as they came off the overflowing stack on my desk.

Thermal Radiation Phenomena, Volume One, Radiative Properties of Air edited by Landshoff, Rolf K. M. and Magee, John L. Plenum Publishing Company, 1969.

The Theory of Heat Radiation, Planck, Max, Dover Publications, 1959.

Faraday as a Discoverer, Tyndall, John, Thomas Y. Crowell

Company, 1961.

Fragments of Science, Tyndall, John, Appleton and Company, 1897.

The New Heat Transfer, Second Edition, Adiutori, Eugene F., Ventuno Press, 1989.

Thermal Radiative Transfer and Properties, Brewster, M. Quinn, Wiley-Interscience, 1992.

Thermal Radiation Heat Transfer, Second Edition, Siegel, Robert and Howell, John R., Hemisphere Publishing, 1981.

Fundamentals of Photonics, Second Edition, Saleh, Bahaa E. A. and Teich, Malvin Carl, Wiley-Interscience, 2007.

Elements of Transport Phenomena, Sissom, Leighton E. and Pitts, Donald R., McGraw-Hill, 1972.

An Introduction to the Kinetic Theory of Gases, Jeans, Sir James, Cambridge University Press, 1946.

Heat Transfer Physics, Kaviany, Massoud, Cambridge University Press, 2008.

Clouds in a Glass of Beer, Simple Experiments in Atmospheric Physics, Bohren, Craig F., Dover Publications, 1987.

Imaging Heat and Mass Transfer Processes, Visualization and Analysis, Panigrahi, Pradipta Kumar and Muralidhar, Krishnamurthy, Springer, 2013.

Design of Thermal Systems, Second Edition, Stoecker, W.F, McGraw-Hill, 1980.

Principles of Heat Transfer, Second Edition, Kreith, Frank, International Textbook Company, 1965.

Heat Transfer, Second Edition, Gebhart, Benjamin, McGraw-Hill, 1971.

A Heat Transfer Textbook, Fourth Edition, Lienhard, John H. IV and Lienhard, John H. V, Dover Publications, 2010.

Radiation Heat Transfer, Augmented Edition, Sparrow, E. M. and Cess, R. D., McGraw-Hill, 1978.

Principles of Planetary Climate, Pierrehumbert, Raymond T., Cambridge University Press, 2010.

A First Course in Atmospheric Thermodynamics, Petty, Grant W., Sundog Publishing, 2008.

A First Course in Atmospheric Radiation, Second Edition, Petty, Grant W., Sundog Publishing, 2006.

More Hot Air, Kordyban, Tony, ASME Press, 2005.

Dynamics of Meteorology and Climate, Scorer, Richard S., Praxis Publishing, 1997.

Heat Transfer Applications for the Practicing Engineer, Theodore, Louis, Wiley-Interscience, 2011.

Weather and Why, Elm, Ienar E., David McKay Company, 1929.

Handbook of Heat Transfer, Rohsenow, Warren M., Hartnett, James P., McGraw-Hill, 1973.

Handbook of Heat Transfer Applications, Second Edition, Rohsenow,

Warren M., Hartnett, James P., Ganic, Ejup N., McGraw-Hill, 1973.

The Hockey Stick and the Climate Wars: Dispatches from the Front Lines, Mann, Michael E., Columbia University Press, 2012.

Radiative Heat Transfer, Love, Tom J., Charles E. Merrill Publishing, 1968.

The Coming Storm: Extreme Weather and our Terrifying Future, Reiss, Bob, Hyperion, 2001.

State of Fear, Crichton, Michael, Harper-Collins, 2004.

An Elementary Treatise on Heat, Fifth Edition, Stewart, Balfour, Clarendon Press, 1888.

The Greenhouse Trap, Daly, John L., Bantam Books, 1989.

www.ingramcontent.com/pod-product-compliance
Lightning Source LLC
Chambersburg PA
CBHW020523270326
41927CB00006B/422